MY LIFE
A RECORD OF EVENTS AND OPINIONS

BY

ALFRED RUSSEL WALLACE

British Library Cataloguing-in-Publication Data
A catalogue record for this book is available from the
British Library

Alfred Russel Wallace

Alfred Russel Wallace was born on 8th January 1823 in the village of Llanbadoc, in Monmouthshire, Wales.

At the age of five, Wallace's family moved to Hertford where he later enrolled at Hertford Grammar School. He was educated there until financial difficulties forced his family to withdraw him in 1836. He then boarded with his older brother John before becoming an apprentice to his eldest brother, William, a surveyor. He worked for William for six years until the business declined due to difficult economic conditions.

After a brief period of unemployment, he was hired as a master at the Collegiate School in Leicester to teach drawing, map-making, and surveying. During this time he met the entomologist Henry Bates who inspired Wallace to begin collecting insects. He and bates continued exchanging letters after Wallace left teaching to pursue his surveying career. They corresponded on prominent works of the time such as Charles Darwin's *The Voyage of the Beagle* (1839) and Robert Chamber's *Vestiges of the Natural History of Creation* (1844).

Wallace was inspired by the travelling naturalists of the day and decided to begin his exploration career collecting specimens in the Amazon rainforest. He explored the Rio Negra for four years, making notes on the peoples and languages he encountered as well as the geography, flora, and fauna. On his return voyage his ship, Helen, caught fire and he and the crew were stranded for ten days before being picked up by the Jordeson, a brig travelling from Cuba to London. All of his specimens aboard Helen had been lost.

After a brief stay in England he embarked on a journey to the Malay Archipelago (now Singapore, Malaysia, and Indonesia). During this eight year period he collected more than 126,000 specimens, several thousand of which represented new species to science. While travelling, Wallace refined his thoughts about evolution and in 1858 he outlined his theory of natural selection in an article he sent to Charles Darwin. This was published in the same year along with Darwin's own theory. Wallace eventually published an account of his travels *The Malay Archipelago* in 1869, and it became one of the most popular books of scientific exploration in the 19th century.

Upon his return to England, in 1862, Wallace became a staunch defender of Darwin's landmark work *On the Origin of Species* (1859). He wrote responses to those critical of the theory of natural selection, including 'Remarks on the Rev. S. Haughton's Paper on the Bee's Cell, And on the Origin of Species' (1863) and 'Creation by Law' (1867). The former of these was particularly pleasing to Darwin. Wallace also published important papers such as 'The Origin of Human Races and the Antiquity of Man Deduced from the Theory of 'Natural Selection" (1864) and books, including the much cited *Darwinism* (1889).

Wallace made a huge contribution to the natural sciences and he will continue to be remembered as one of the key figures in the development of evolutionary theory.

Wallace died on 7th November 1913 at the age of 90. He is buried in a small cemetery at Broadstone, Dorset, England.

PREFACE TO THE NEW EDITION

AMONG the numerous kind and even flattering notices of the first edition of this work, there were a considerable number in which objection was made to its great bulk, caused in part by the inclusion in it of subjects only indirectly related to myself, as well as of some of my early writings which were of no special interest.

Recognizing as I do the justice of this criticism, I gladly agreed to the suggestion of my publishers that I should prepare a new edition in one volume, by omitting all such superfluous matter as is above referred to.

Feeling that I was not myself the best judge of what to omit, I asked my son, Mr. William G. Wallace, to undertake this task, after agreeing with him, and with my publishers, on several entire chapters which must certainly be omitted. In order to represent the general reader he asked a friend of his own, who had not read the book before, to assist him in forming an opinion in doubtful cases. I have also myself condensed some diffuse portions, and have added a few additional facts, bringing the story of "My Life" down to the present time.

All the illustrations have been retained which are in any way referred to in the present work, and I trust that the result will be to render it acceptable to a new body of readers.

BROADSTONE,

October 1, 1908.

PREFACE TO THE ORIGINAL EDITION

THE present volumes would not have been written had not the representatives of my English and American publishers assured me that they would probably interest a large number of readers.

I had indeed promised to write some account of my early life for the information of my son and daughter, but this would have been of very limited scope, and would probably not have been printed.

Having never kept a diary, except when abroad, nor preserved any of the earlier letters of my friends, I at first thought that I had no materials for any full record of my life and experiences. But when I set to work in earnest to get together whatever scattered memoranda I could find, the numerous letters I possessed from men of considerable eminence, dating from my return home in 1862, together with a few of my own returned to me by some of my correspondents, I began to see that I had a fair amount of material, though I was very doubtful how far it would interest any considerable number of readers.

As several of my friends have assured me that a true record of a life, especially if sufficiently full as to illustrate development of character so far as that is due to environment, would be extremely interesting, I have kept this in mind, perhaps unduly, though I am not at all sure that my own conclusions on this point are correct.

It is difficult to write such a record as mine (extending to the memories of nearly eighty years) without subjecting one's self to the charge of diffuseness or egotism, and I cannot hope to escape this altogether. But as my experiences have been certainly varied, if not exciting, I trust that the frequent change of scene and of occupation, together with the diversity of my interests and of the persons with whom I have been associated, may render this story of my life less tedious than might have been anticipated.

My thanks are due to those friends who have assisted me with facts or illustrations, and especially to Mrs. Arthur Waugh, who has been so kind as to make the very full Index to my book.

OLD ORCHARD, BROADSTONE,
September, 1905.

CONTENTS

CONTENTS

CONTENTS

CONTENTS

MY LIFE

A RECORD OF EVENTS AND

OPINIONS

CHAPTER I

MY RELATIVES AND ANCESTORS

OUR family had but few relations, and I myself never saw a grandfather or grandmother, nor a true uncle, and but one aunt—my mother's only sister. The only cousins we ever had, so far as I know, were that sister's family of eight or nine, all but two of whom emigrated to South Australia in 1838. Of the two who remained in England, the daughter had married Mr. Burningham, and had only one child, a daughter, who has never married. The son, the Rev. Percy Wilson, had a family, none of whom, however, I have ever met, though I have recently had a visit from a son of another cousin, Algernon, with whom I had a considerable correspondence.

My father was practically an only son, an elder boy dying when three months old; and as his father died when he was a boy of twelve, and his mother when he was an infant, he had not much opportunity of hearing about the family history. I myself left home before I was fourteen, and only rarely visited my parents for short holidays, except once during my recovery from a dangerous illness, so that I also had little opportunity of learning anything of our ancestors on the paternal side, more especially as my father seldom spoke of his youth, and I as a boy felt no interest in his genealogy. Neither did my eldest brother William—with whom I lived till I was of age —ever speak on the subject. The little I have gleaned was from my sister Fanny and from a recent examination of tombstones and parish registers, and especially from an old Prayer-book (1723) which belonged to my grandfather Wallace, who had registered in it the dates of the births and baptisms of his two sons, while my father had continued the register to include his own family of nine children, of whom I am the only survivor.

My paternal grandfather was married at Hanworth, Middlesex, in 1765, and the parish register describes him as William Wallace, of Hanworth, bachelor, and his wife as Elizabeth Dilke, of Laleham, widow. Both are buried in Laleham churchyard, where I presume the former Mrs. Dilke had some family burial rights, as my grandfather's brother, George Wallace, is also buried there. The register at Hanworth contains no record of my father's birth, but the church itself shows that quite a small colony of Wallaces lived at Hanworth. On a long stone in the floor of the chancel is the name of JAMES WALLACE, ESQ., who died February 7, 1778, aged 87 years. He was therefore thirty-five years older than my grandfather, and may have been his uncle. Then follows ADMIRAL SIR JAMES WALLACE, who died on March 6, 1803, aged 69 years; and FRANCES SLEIGH, daughter of the above JAMES WALLACE, ESQ., who died December 12, 1820, aged 69 years.

How or why my grandfather came to live at Hanworth (probably with his brother George, who is also buried at Laleham), I can only conjecture from the following facts. Baron Vere of Hanworth is one of the titles of the Dukes of St. Albans since 1750, when Vere Beauclerc, third son of the first Duke, was created Baron, and his son became fifth Duke of St. Albans in 1787. It is to be presumed that the village and a good deal of the land was at that time the property of this family, though they appear to have parted with it not long afterwards, as a Mr. Perkins owned the park and rebuilt the church in 1812. The St. Albans family have a tomb in the church. Now, my father's name was Thomas

Vere Wallace, and it therefore seems probable that his father was a tenant of the first Baron Vere, and in his will he is styled "Victualler." He probably kept the inn on the estate.

The only further scrap of information as to my father's family is derived from a remark he once made in my hearing, that his uncles at Stirling (I think he said) were very tall men. I myself was six feet in height when I was sixteen, and my eldest brother William was an inch taller, while my brother John and sister Fanny were both rather tall. My father and mother, however, were under rather than over middle height, and the remark about his tall uncles was to account for this abnormal height by showing that it was in the family. As all the Wallaces of Scotland are held to be various branches of the one family of the hero Sir William Wallace, we have always considered ourselves to be descended from that famous stock; and this view is supported by the fact that our family crest was said to be an ostrich's head with a horseshoe in its mouth, and this crest belongs, according to Burke's "Peerage," to Craigie-Wallace, one of the branches of the patriot's family.

Of my mother's family I have somewhat fuller details, though not going any further back. Her father was John Greenell, of Hertford, who died there in 1824, at the age of 79. He had two daughters, Martha and Mary Anne, my mother. Their mother died when the two girls were two and three years old. Mr. Greenell married a second time, and his widow lived till 1828. My mother's grandfather, who died in 1797, aged 80, was for many years an alderman, and twice Mayor of Hertford (in 1773 and 1779), as stated in the records of the borough. He was buried in St. Andrew's churchyard.

There is also in the same churchyard a family tomb, in which my father and my sister Eliza are buried, but which belonged to a brother of my mother's grandfather, William Greenell, as shown by the inscription.

I will now say a few words about my father's early life, and the various family troubles which, though apparently very disadvantageous to his children, may yet have been on the whole, as is so often the case, benefits in disguise.

My father, Thomas Vere Wallace, was twelve years old when his father died, but his stepmother lived twenty-one years after her husband, and I think it not improbable that she may have resided in Marylebone near William Greenell the architect, and that my father went to school there. The only thing I remember his telling us about his school was that his master dressed in the old fashion, and that he had a best suit entirely of yellow velvet.

When my father left school he was articled to a firm of solicitors—Messrs. Ewington and Chilcot, Bond Court, Walbrook, I think, as I find this name in an old note-book of my father's—and in 1792, when he had just come of age, he was duly sworn in as an Attorney-at-Law of the Court of

King's Bench. He is described in the deed of admittance as of Lamb's Conduit Street, where he probably lodged while pursuing his legal studies, it being near the Inns of Court and at the same time almost in the country. He seems, however, never to have practised law, since he came into property which gave him an income of about £500 a year.

From this time till he married, fifteen years later, he appears to have lived quite idly, so far as being without any systematic occupation, often going to Bath in the season, where, he used to tell us, he had met the celebrated Beau Brummell and other characters of the early years of the nineteenth century.

My sister told me that while he remained a bachelor my father lived up to his income or very nearly so; and from what we know of his after life this did not imply any extravagance or luxurious habits, but simply that he enjoyed himself in London and the country, living at the best inns or boarding-houses, and taking part in the amusements of the period, as a fairly well-to-do, middle-class gentleman.

After his marriage in 1807 he lived in Marylebone, and his ordinary household expenses, of course, increased; and as by 1810 he had two children and the prospect of a large family, he appears to have felt the necessity of increasing his income. Having neglected the law so long, and probably having a distaste for it, he apparently thought it quite hopeless to begin to practise as a solicitor, and being entirely devoid of business habits, allowed himself to be persuaded into undertaking one of the most risky of literary speculations, the starting a new illustrated magazine, devoted apparently to art, antiquities, and general literature. After a very few numbers were issued the whole thing came to grief, partly, it was said, by the defalcations of a manager or book-keeper, who appropriated the money advanced by my father to pay for work and materials, and partly, no doubt, from the affair being in the hands of persons without the necessary business experience and literary capacity to make it a success.

The result was that my father had to bear almost the whole loss, and this considerably reduced his already too scanty income. Whether he made any other efforts to earn money I do not know, but he continued to live in Marylebone till 1816, a daughter Emma having been born there in that year; but soon after he appears to have removed to St. George's, Southwark, in which parish my brother John was born in 1818. Shortly afterwards his affairs must have been getting worse, and he determined to move with his family of six children to some place where living was as cheap as possible; and, probably from having introductions to some residents there, fixed upon Usk, in Monmouthshire, where a sufficiently roomy cottage with a large garden was obtained, and where I was born on January 8, 1823.

In the year 1828 my mother's step-mother, Mrs. Rebecca Greenell, died at Hertford, and I presume it was in consequence of this event that the family left Usk in that year, and lived at Hertford for the next nine or ten years, removing to Hoddesdon in 1837 or 1838, where my father died in 1843. These last fifteen years of his life were a period of great trouble and anxiety, his affairs becoming more and more involved, till at last the family became almost wholly dependent on my mother's small marriage settlement of less than a hundred a year, supplemented by his taking a few pupils and by a small salary which he received as librarian to a subscription library.

During the latter part of the time we lived at Hertford his troubles were great. He appears to have allowed a solicitor and friend whom he trusted to realize what remained of his property and invest it in ground-rents which would bring in a larger income, and at the same time be perfectly secure. For a few years the income from this property was duly paid him, then it was partially and afterwards wholly stopped. It appeared that the solicitor was himself engaged in a large building speculation in London, which was certain to be ultimately of great value, but which he had not capital enough to complete. He therefore had to raise money, and did so by using funds entrusted to him for other purposes, among them my father's small capital. But, unfortunately, other creditors pressed upon him, and he was obliged to sacrifice the whole of the building estate at almost a nominal price. Out of the wreck of the solicitor's fortune my father obtained a small portion of the money due, with promises to pay all at some future time. Among the property thus lost were some legacies from my mother's relations to her children, and the whole affair got into the hands of the lawyers, from whom small amounts were periodically received which helped to provide us with bare necessaries.

As a result of this series of misfortunes the children who reached their majority had little or nothing to start with in earning their own living, except a very-ordinary education, and a more or less efficient training. The eldest son, William, was first articled to a firm of surveyors at Kington, Herefordshire, probably during the time we resided at Usk. He then spent a year or two in the office of an architect at Hertford, and finally a year in London with a large builder named Martin then engaged in the erection of King's College, in order to became familiar with the practical details of building. He may be said, therefore, to have had a really good professional education. At first he got into general land-surveying work, which was at that time rather abundant, owing to the surveys and valuations required for carrying out the Commutation of Tithes Act of 1836, and also for the enclosures of commons which were then very frequent. During the time I was with him we were largely engaged in this kind of work in various parts of England and Wales, as will be seen later on; but the payment for such work was by no means liberal, and owing to the frequent periods of idleness between one job and another, it was about as much as my brother could do to earn our living and travelling expenses.

About the time I went to live with my brother (1837) my sister Fanny entered a French school at Lille to learn the language and to teach English, and I think she was a year there. On her return she started the school at Hoddesdon, but after my father's death in 1843, she obtained a position as a teacher in Columbia College, Georgia, U.S.A., then just established under the Bishop of Georgia; and she only returned after my brother William's death in 1846, when the surviving members of the family in England were reunited, and lived together for two years in a cottage near Neath, in Glamorganshire.

My brother John, at the age of fourteen or fifteen, was apprenticed, first to Mr. Martin and then to Mr. Webster, a Londer builder living in Albany Street, Regent Park, where he became a thorough joiner and carpenter. He afterwards worked for a time for Cubitt and other large builders; then, when he came to live with me at Neath, he learnt surveying and a little architecture. When I went to the Amazon, he took a small dairy-farm at too high a rent, and not making this pay, in 1849 he emigrated to California at the height of the first rush for gold, joined several mining camps, and was moderately successful. About five years later he came home, married Miss Webster, and returning to California, settled for some years at Columbia, a small mining town in Tuolumne County. He afterwards removed to Stockton, where he practised as surveyor and water engineer till his death in 1895.

My younger brother, Herbert, was first placed with a trunk maker in Regent Street, but not liking this business, afterwards came to Neath and entered the pattern-shops of the Neath Ironworks. After his brother John went to California he came out to me at Para, and after a year spent on the Amazon as far as Barra on the Rio Negro, he returned to Para on his way home, where he caught yellow fever, and died in a few days at the early age of twenty-two. He was the only member of our family who had a considerable gift of poesy, and was probably more fitted for a literary career than for any mechanical or professional occupation.

It will thus be seen that we were all of us very much thrown on our own resources to make our way in life; and as we all, I think, inherited from my father a certain amount of constitutional inactivity or laziness, the necessity for work that our circumstances entailed was certainly beneficial in developing whatever powers were latent in us; and this is what I implied when I remarked that our father's loss of his property was perhaps a blessing in disguise.

Of the five daughters, the first-born died when five months old; the next, Eliza, died of consumption at Hertford, aged twenty-two. Two others, Mary Anne and Emma, died at Usk at the ages of eight and six respectively; while Frances married Mr. Thomas Sims, a photographer, and died in London, aged eighty-one.

On the whole, both the Wallaces and the Greenells seem to have been rather long-lived when they reached manhood or womanhood. The five ancestral Wallaces of whom I have records had an average age of seventy years, while the five Greenells had an average of seventy-six. Of our own family, my brother John reached seventy-seven, and my sister Fanny eighty-one. My brother William owed his death to a railway journey by night in winter, from London to South Wales in the miserable accommodation then afforded to third-class passengers, which, increased by a damp bed at Bristol, brought on severe congestion of the lungs, from which he never recovered.

I will now give a short account of my father's appearance and character. In a miniature, painted just before his marriage, when he was thirty-five years old, he is represented in a blue coat with gilt buttons, a white waistcoat, a thick white neck-cloth coming up to the chin and showing no collar, and a frilled shirtfront. This was probably his wedding-coat, and his usual costume, indicating the transition from the richly coloured semi-court dress of the earlier Georgian period to the plain black of our own day. He is shown as having a ruddy complexion, blue eyes, and carefully dressed and curled hair, which I think must have been powdered, or else in the transition from light brown to pure white. As I remember him from the age of fifty-five onwards, his hair was rather thin and quite white, and he was always clean-shaved as in the miniature.

THOMAS VERE WALLACE. AGED 36.
(At time of his marriage.)

MARY ANNE WALLACE. AGED 18.
(At time of her marriage.)

In figure he was somewhat below the middle height. He was fairly active and fond of gardening and other country occupations, such as brewing beer and making grape or elder wine whenever he had the opportunity; and during some years at Hertford he rented a garden about half a mile away, in order to grow vegetables and have some wholesome exercise.

He was rather precise and regular in his habits, quiet and rather dignified in manners, and what would be termed a gentleman of the old school. Of course, he always wore a top-hat—a beaver hat as it was then called, before silk hats were invented—the only other headgear being sometimes a straw hat for use in the garden in summer.

In character he was quiet and even-tempered, very religious in the orthodox Church-of-England way, and with such a reliance on Providence as almost to amount to fatalism. He was fond of reading, and through reading-clubs or lending-libraries we usually had some of the best books of travel or biography in the house. Some of these my father would read to us in the evening, and when Bowdler's edition of Shakespeare came out he obtained it, and often read a play to the assembled family.

At one time my father wrote a good deal, and we were told it was a history of Hertford, or at other times some religious work; but they were never finished, and I do not think they would ever have been worth publishing, his character not leading him to do any such work with sufficient thoroughness. He dabbled a little in antiquities and in heraldry, but did nothing systematic; and though he had fair mental ability he possessed no special talent, either literary, artistic, or scientific. He sketched a little, but with a very weak and uncertain touch, and among his few scrap- and note-books that have been preserved, there is hardly anything original except one or two short poems in the usual didactic style of the period, but of no special merit.

CHAPTER II

MY EARLIEST MEMORIES, USK AND HERTFORD

MY earliest recollections of myself are as a little boy in short frocks and with bare arms and legs, playing with my brother and sisters, or sitting in my mother's lap or on a footstool listening to stories, of which some fairy-tales, especially "Jack the Giant-Killer," "Little Red Riding Hood," and "Jack and the Beanstalk," seem to live in my memory; and of a more realistic kind, "Sandford and Merton," which perhaps impressed me even more deeply than any. I clearly remember the little house and the room we chiefly occupied, with a French window opening to the garden, a steep wooded bank on the right, the road, river, and distant low hills to the left. The house itself was built close under this bank, which was quite rocky in places.

The river in front of our house was the Usk, a fine stream on which we often saw men fishing in coracles, the ancient form of boat made of strong wicker-work, somewhat the shape of the deeper half of a cockleshell, and covered with bullock's hide. Each coracle held one man, and it could be easily carried to and from the river on the owner's back. In those days of scanty population and abundant fish the river was not preserved, and a number of men got their living, or part of it, by supplying the towns with salmon and trout in their season. It is very interesting that this extremely ancient boat, which has been in use from pre-Roman times, and perhaps even from the Neolithic Age, should continue to be used on several of the Welsh rivers down to the present day. There is probably no other type of vessel now in existence which has remained unchanged for so long a period.

The chief attraction of the river to us children was the opportunity it afforded us for catching small fish, especially lampreys. A short distance from our house, towards the little village of Llanbadock, the rocky bank came close to the road, and a stone quarry had been opened to obtain stone, both for building and road-mending purposes. Here, occasionally, the rock was blasted, and sometimes we had the fearful delight of watching the explosion from a safe distance, and seeing a cloud of the smaller stones shoot up into the air. At some earlier period very large charges of powder must have been used, hurling great slabs of rock across the road into the river, where they lay, forming convenient piers and standing-places on its margin. Some of these slabs were eight or ten feet long and nearly as wide; and it was these that formed our favourite fishing-stations, where we sometimes found shoals of small lampreys, which could be scooped up in basins or old saucepans, and were then fried for our dinner or supper, to our great enjoyment. I think what we caught must have been the young fish, as my recollection of them is that they were like little eels, and not more than six or eight inches long, whereas the full-grown lampreys are from a foot and a half to nearly three feet long.

At this time I must have been about four years old, as we left Usk when I was about five, or less. My brother John was four and a half years older, and I expect was the leader in most of our games and explorations. My two sisters were five and seven years older than John, so that they would have been about thirteen and fifteen, which would appear to me quite grown up; and this makes me think that my recollections must go back to the time when I was just over three, as I quite distinctly remember two, if not three, besides myself, standing on the flat stones and catching lampreys.

There is an incident in which I remember that my brother and at least one, if not two, of my sisters took part. Among the books read to us was "Sandford and Merton," the only part of which that I distinctly remember is when the two boys got lost in a wood after dark, and while Merton could do nothing but cry at the idea of having to pass the night without supper or bed, the resourceful Sandford comforted him by promising that he should have both, and set him to gather sticks for a fire, which he lit with a tinder-box and match from his pocket. Then, when a large fire had been made, he produced some potatoes which he had picked up in a field on the way, and which he then roasted beautifully in the embers, and even produced from another pocket a pinch of salt in a screw of paper, so that the two boys had a very good supper. Then, collecting fern and dead leaves for a bed, and I think making a coverlet by taking off their two jackets, which made them quite comfortable while lying as close together as possible, they enjoyed a good night's sleep till daybreak, when they easily found their way home.

This seemed so delightful that one day John provided himself with the matchbox, salt, and potatoes, and having climbed up the steep bank behind our house, as we often did, and passed over a field or two to the woods beyond, to my great delight a fire was made, and we also feasted on potatoes with salt, as Sandford and Merton had done. Of course we did not complete the imitation of the story by sleeping in the wood, which would have been too bold and dangerous an undertaking for our sisters to join in, even if my brother and I had wished to do so.

I may here mention a psychological peculiarity, no doubt common to many children of the same age, that, during the whole period of my residence at Usk, I have no clear recollection, and can form no distinct mental image, of either my father or mother, brothers or sisters. I simply recollect that they existed, but my recollection is only a blurred image, and does not extend to any peculiarities of feature, form, or even of dress or habits. It is only at a considerably later period that I begin to recollect them as distinct and well-marked individuals whose form and features could not be mistaken—as, in fact, being *my* father and mother, my brothers and sisters; and the house and surroundings in which I can thus first recollect, and in some degree visualize them, enable me to say that I must have been then at least eight years old.

What makes this deficiency the more curious is that, during the very same period at which I cannot recall the personal appearance of the individuals with whom my life was most closely associated, I *can* recall all the main features and many of the details of my outdoor, and, to a less degree, of my indoor, surroundings. The form and colour of the house, the road, the river close below it, the bridge with the cottage near its foot, the narrow fields between us and the bridge, the steep wooded bank at the back, the stone quarry and the very shape and position of the flat slabs on which we stood fishing, the cottages a little further on the road, the little church of Llanbadock and the stone stile into the churchyard, the fishermen and their coracles, the ruined castle, its winding stair and the delightful walk round its top—all come before me as I recall these earlier days with a distinctness strangely contrasted with the vague shadowy figures of the human beings who were my constant associates in all these scenes. In the house I recollect the arrangement of the rooms, the French window to the garden, and the blue-papered room n which I slept, but of the people always with me in those rooms, and even of the daily routine of our life, I remember nothing at all.

I cannot find any clear explanation of these facts in modern psychology, whereas they all become intelligible from the phrenological point of view. The shape of my head shows that I have *form* and *individuality* but moderately developed, while *locality, ideality, colour*, and *comparison* are decidedly stronger. Deficiency in the first two caused me to take little notice of the characteristic form and features of the separate individualities which were most familiar to me, and from that very cause attracted less close attention; while the greater activity of the latter group gave interest and attractiveness to the ever-changing combinations in outdoor scenery, while the varied opportunities for the exercise of the physical activities, and the delight in the endless variety of nature which are so strong in early childhood, impressed these outdoor scenes and interests upon my memory. And throughout life the same limitations of observation and memory have been manifest. In a new locality it takes me a considerable time before I learn to recognize my various new acquaintances individually; and looking back on the varied scenes amid which I have lived at home and abroad, while numerous objects, localities, and events are recalled with some distinctness, the people I met, or, with few exceptions, those with whom I became fairly well acquainted, seem but blurred and indistinct images.

In the year 1883, when for the first time since my childhood I revisited, with my wife and two children, the scenes of my infancy, I obtained a striking proof of the accuracy of my memory of those scenes and objects. Although the town of Usk had grown considerably on the north side towards the railway, yet, to my surprise and delight, I found that no change whatever had occurred on our side of the river, where, between the bridge and Llanbadock, not a new house had been built, and our cottage and garden, the path up to the front door, and the steep woody bank behind it, remained exactly as pictured in my memory. Even the quarry appeared to have been very little enlarged, and the great flat stones were still in the river exactly as when I had stood upon them with my brothers and sisters sixty years before. The one change I noted here was that the well-remembered stone stile into the village churchyard had been replaced by a wooden one. We also visited the ruined castle, ascended the winding stair, and walked round the top wall, and everything seemed to me exactly as I knew it of old, and neither smaller nor larger than my memory had so long pictured it. The view of the Abergavenny mountains pleased and interested me as in childhood, and the clear-flowing Usk seemed just as broad and as pleasant to the eye as my memory had always pictured it.

So far as I can remember or have heard I had no illness

of any kind at Usk, which was no doubt due to the free outdoor life we lived there, spending a great part of the day in the large garden or by the riverside, or in the fields and woods around us. As will be seen later on, this immunity ceased as soon as we went to live *in* a town. I remember only one childish accident. The cook was taking away a frying-pan with a good deal of boiling fat in it, which for some reason I wanted to see, and, stretching out my arm over it, I suppose to show that I wanted it lowered down, my fore-arm went into the fat and was badly scalded. I mention this only for the purpose of calling attention to the fact that, although I vividly remember the incident, I cannot recall that I suffered the least pain, though I was told afterwards that it was really a severe burn. This, and other facts of a similar kind, make me think that young children suffer far less pain than adults from the same injuries. And this is quite in accordance with the purpose for which pain exists, which is to guard the body against injuries dangerous to life, and to give us the impulse to escape rapidly from any danger. But as infants cannot escape from fatal dangers, and do not even know what things are dangerous and what not, only very slight sensations of pain are at first required, and such only are therefore developed, and these increase in intensity just in proportion as command over the muscles giving the power of rapid automatic movements becomes possible. The sensation of pain does not, probably, reach its maximum till the whole organism is fully developed in the adult individual. This is rather a comforting conclusion in view of the sufferings of so many infants needlessly massacred through the terrible defects of our vicious social system.

I may add here a note as to my personal appearance at this age. I was exceedingly fair, and my long hair was of a very light flaxen tint, so that I was generally spoken of among the Welsh-speaking country people as the "little Saxon."

My recollections of our leaving Usk and of the journey to London are very faint, only one incident of it being clearly visualized—the crossing the Severn at the Old Passage in an open ferry-boat. This is very clear to me, possibly because it was the first time I had ever been in a boat. I remember sitting with my mother and sisters on a seat at one side of the boat, which seemed to me about as wide as a small room, of its leaning over so that we were close to the water, and especially of the great boom of the mainsail, when our course was changed, requiring us all to stoop our heads for it to swing over us. It was a little awful to me, and I think we were all glad when it was over and we were safe on land again. We must have travelled all day by coach from Usk to the Severn, then on to Bristol, then from Bristol to London, where we stayed at an inn in Holborn.

Of the next few months of my life I have also but slight recollections, confined to some isolated facts or incidents. On leaving the inn we went to my aunt's at Dulwich. I remember being much impressed with the large house, and especially with the beautiful grounds, with lawns, trees, and shrubs such as I had never seen before. There was here also a family of cousins, some about my own age, and the few days we stayed were very bright and enjoyable. Thence I was sent to a school in Essex, with some ladies who had about half a dozen boys and girls of my own age.

My next recollections are of the town of Hertford, where we lived for eight or nine years, and where I received the whole of my school education. We had a small house, the first of a row of four at the beginning of St. Andrew's Street; and I must have been a little more than six years old when I first remember myself in this house, which had a very narrow yard at the back, and a dwarf wall, perhaps five feet high, between us and the adjoining house. The very first incident I remember, which happened, I think, on the morning after my arrival, was of a boy about my own age looking over this wall, who at once inquired, "Hullo! who are you?" I told him that I had just come, and what my name was, and we at once made friends. The stand of a water-butt enabled me to get up and sit upon the wall, and by means of some similar convenience he could do the same, and we were thus able to sit side by side and talk, or get over the wall and play together when we liked. Thus began the friendship of George Silk and Alfred Wallace, which, with long intervals of absence at various periods, has continued to this day.

The old town of Hertford, in which I passed the most impressionable years of my life, and where I first obtained a rudimentary acquaintance with my fellow-creatures and with nature, is, perhaps, on the whole, one of the most pleasantly situated county towns in England, although as a boy I did not know this, and did not appreciate the many advantages I enjoyed. Among its most delightful features are numerous rivers and streams in the immediately surrounding country, affording pleasant walks through flowery meads, many picturesque old mills, and a great variety of landscape. The river Lea, coming from the south-west, passes through the middle of the town, where the old town-mill was situated in an open space called the Wash, which was no doubt liable to be flooded in early times. The miller was reputed to be one of the richest men in the town, yet we often saw him standing at the mill doors in his dusty miller's clothes as we passed on our way from school. He was a cousin of my mother's by marriage, and we children sometimes went to tea at his house, and then, as a great treat, were shown all over the mill with all its strange wheels and whirling millstones, its queer little pockets, on moving leather belts, carrying the wheat up to the stones in a continual stream, the ever-rattling sieves and cloths which sifted out the bran and pollard, and the weird peep into the dark cavern where the great dripping water-wheel went on its perpetual round. Where the river passed under the bridge close by, we could clamber up and look over the parapet into the deep, clear water rushing over a dam, and also see where the stream that turned the wheel passed swiftly under a low arch, and this was a sight that never palled upon us, so that almost every fine day, as we passed this way home from school, we gave a few moments to gazing into this dark, deep water, almost always in shadow owing to high buildings on both sides of it, but affording a pleasant peep to fields and gardens beyond.

Here, too, in the river Beane, which had a gentle stream with alternate deep holes and sandy shallows, suitable for boys of all ages, was our favourite bathing-place, where, not long after our coming to Hertford, I was very nearly drowned. It was at a place called Willowhole, where those who could swim a little would jump in, and in a few strokes in any direction reach shallower water. I and my brother John and several schoolfellows were going to bathe, and I, who had undressed first, was standing on the brink, when one of my companions gave me a sudden push from behind, and I tumbled in and went under water immediately. Coming to the surface half dazed, I splashed about and went under again, when my brother, who was four and a half years older, jumped in and pulled me out. I do not think I had actually lost consciousness, but I had swallowed a good deal of water, and I lay on the grass for some time before I got strength to dress, and by the time I got home I was quite well. It was, I think, the first year, if not the first time, I had ever bathed, and if my brother had not been there it is quite possible that I might have been drowned. This gave me such a fright that though I often bathed here afterwards, I always went in where the water was shallow, and did not learn to swim, however little, till several years later.

Few small towns (it had then less than six thousand inhabitants) have a more agreeable public playground than the fine open space called Hartham with the level valley of the Lea stretching away to Ware on the east, the town itself just over the river on the south, while on the north, just across the river Beane, was a steep slope covered with scattered fir trees, and called the Warren, at the foot of which was a footpath leading to the picturesque little village and old church of Bengeo. This path along the Warren was a favourite walk of mine either alone or with a playmate, where we could scramble up the bank, climb up some of the old trees, or sit comfortably upon one or two old stumpy yews, which had such twisted branches and stiff spreading foliage as to form delightful seats. This place was very little frequented, and our wanderings in it were never interfered with.

About three-quarters of a mile from the centre of the town, going along West Street, was a mill called Horn's Mill, which was a great attraction to me. It was an old-fashioned mill for grinding linseed, expressing the oil, and making oil-cake. The mill stood close by the roadside, and there were small low windows always open, through which we could look in at the fascinating processes as long as we liked. First, there were two great vertical millstones of very smooth red granite, which shone beautifully from the oil of the ground seeds. These were fixed on each side of a massive vertical wooden axis on a central iron axle, revolving slowly and silently, and crushing the linseed into a fine oily meal.

But this was only one part of this delightful kind of peepshow. A little way off an equally novel and still more complex operation was always going on, accompanied by strange noises always dear to the young. Looking in at other windows we saw numbers of workmen engaged in strange operations amid strange machinery, with its hum and whirl and reverberating noises. Close before us were long erections like shop counters, but not quite so high. Immediately above these, at a height of perhaps ten or twelve feet, a long cylindrical beam was continually revolving with fixed beams on each side of it, both higher up and lower down. At regular intervals along the counter were great upright wooden stampers shod with iron at the bottom. When not in action these were supported so that they were about two feet above the counter, and just below them was a square hole. As we looked on a man would take a small canvas sack about two feet long, fill it full of linseed meal from a large box by his side, place this bag in a strong cover of a kind of floorcloth with flaps going over the top and down each side.

The sack of meal thus prepared would then be dropped into the hole, which it entered easily. Then a thin board of hard wood, tapered to the lower edge, was pushed down on one side of it, and outside this again another wedge-shaped piece was inserted. The top of this was now just under the iron cap of the heavy pile or rammer, and on pulling a rope, this was freed and dropped on the top of the wedge, which it forced halfway down. In a few seconds it was raised up again, and fell upon the wedge, driving it in a good deal further, and the third blow would send it down level with the top of the counter. Then when the rammer rose up, another rope was pulled, and it remained suspended; a turn of a handle enabled the first wedge to be drawn out and a much thicker one inserted, when, after two or three blows, this became so hard to drive that the rammer falling upon it made a dull sound and rebounded a little; and as the process went on the blows became sharper, and the pile would rebound two or three times like a billiard ball rebounding again and again from a stone floor, but in more rapid succession. This went on for hours, and when the process was finished, the meal in the sack had become so highly compressed that when taken out it was found to be converted into a compact oilcake.

CHAPTER III

MY SCHOOL LIFE AT HERTFORD

MY recollections of life at our first house in St. Andrew's Street are very scanty. My father had about half a dozen small boys to teach, and we used to play together; but I think that when we had been there about a year or two, I went to the Grammar School with my brother John, and was at once set upon that most wearisome of tasks, the Latin grammar. It was soon after this that I had the first of the three serious illnesses which at different periods brought me within a few hours of death in the opinion of those around me.

It was a severe attack of scarlet fever, and I remember little but heat and horrid dreams till one evening when all the family came to look at me, and I had something given me to drink all night. I was told afterwards that the doctor said this was the crisis, that I was to have port wine in teaspoonfuls at short intervals, and that if I was not dead before morning I might recover. For some weeks after this I lived a very enjoyable life in bed, having tea and toast, puddings, grapes, and other luxuries till I was well again.

It may be well here to give a brief outline of my school life at Hertford and of the schoolmaster who taught me. The school itself was built in the year 1617, when the school was founded. It consisted of one large room, with a large square window at each end and two on each side. In the centre of one side was a roomy porch, and opposite to it a projecting portion, with a staircase leading to two rooms above the schoolroom and partly in the roof. The schoolroom was fairly lofty. Along the sides were what were termed porches—desks and seats against the wall with very solid, roughly carved ends of black oak, much cut with the initials or names of many generations of schoolboys. In the central space were two rows of desks with forms on each side. There was a master's desk at each end, and two others on the sides, and two open fireplaces equidistant from the ends. Every boy had a desk the sloping lid of which opened, to keep his school-books and anything else he liked, and between each pair of desks at the top was a leaden ink-pot, sunk in a hole in the middle rail of the desks. As we went to school even in winter at seven in the morning, and three days a week remained till five in the afternoon, some artificial lighting was necessary, and this was effected by the primitive method of every boy bringing his own candles or candle-ends with any kind of candlestick he liked. An empty ink-bottle was often used, or the candle was even stuck on to the desk with a little of its own grease. So that it enabled us to learn our lessons or do our sums, no one seemed to trouble about how we provided the light.

The school was reached by a path along the bottom of All Saints' Churchyard, and entered by a door in the wall which entirely surrounded the school playground and master's garden. Over this door was a Latin motto—

"Inter umbras Academi studere delectat."

This was appropriate, as the grounds were surrounded by trees, and at the north end of the main playground there were two very fine elms, shown in the old engraving of the school here reproduced.

The headmaster in my time was a rather irascible little man named Clement Henry Crutwell. He was usually called by the boys Old Cruttle or Old Clemmy, and when he overheard these names used, which was not often, he would give us a short lecture on the impropriety and impoliteness of miscalling those in authority over us. He was a good master, inasmuch as he kept order in the school, and carried on the work of teaching about eighty boys by four masters, all in one room, with great regularity and with no marked inconvenience. Whatever might be the noise and games going on when he was absent, the moment his step was heard in the porch silence and order at once reigned.

Flogging with a cane was not uncommon for more serious offences, while for slighter ones he would box the ears severely. Caning was performed in the usual old-fashioned way by laying the boy across the desk, his hands being held on one side and his feet on the other, while the master, pulling the boy's trousers tight with one hand, laid on the cane with great vigour with the other. Mr. Crutwell always caned the boys himself, but the other masters administered minor punishments, such as slight ear-boxes, slapping the palm with a flat ruler, or rapping the knuckles with a round one. These punishments were usually deserved, though not always. A stupid boy, or one who had a bad verbal memory, was often punished for what was called invincible idleness when it was really congenital incapacity to learn what he took no interest in, or what often had no meaning for him.

Mr. Crutwell was, I suppose, a fairly good classical scholar, as he took the higher classes in Latin and Greek. I left school too young even to begin Greek, but the last year or two I was in the Latin class which was going through Virgil's "æneid" with him. The system was very bad. The eight or ten boys in the class had an hour to prepare the translation, and they all sat together in a group opposite each other and close to Mr. Crutwell's desk, but under pretence of work there were always two or three of the boys who were full of talk and gossip and school stories, which kept us all employed and amused till within about a quarter of an hour of the time for being called up, when some one would remark, "I say, let's do our translation; I don't know a word of it." Then the cleverest boy, or one who had already been through the book, would begin to translate, two or three others would have their dictionaries ready when he did not know the

THE GRAMMAR SCHOOL, HERTFORD.

meaning of a word, and so we blundered through our forty or fifty lines. When we were called up, it was all a matter of chance whether we got through well or otherwise. If the master was in a good humour and the part we had to translate was specially interesting, he would help us on whenever we hesitated or blundered, and when we had got through the lesson, he would make a few remarks on the subject, and say, "Now I will read you the whole incident." He would then take out a translation of the "æneid" in verse by a relative of his own—an uncle, I think—and, beginning perhaps a page or two back, read us several pages, so that we could better appreciate what we had been trying to translate. I, for one, always enjoyed these readings, as the verse was clear and melodious, and gave an excellent idea of the poetry of the Latin writer. Sometimes our laziness and ignorance were found out, and we either had to stay in an hour and go over it again, or copy it out a dozen times, or some other stupid imposition. But as this only occurred now and then, of course it did not in the least affect our general mode of procedure when supposed to be learning our lesson. Mr. Crutwell read well, with a good emphasis and intonation, and I obtained a better idea of what Virgil really was from his readings than from the fragmentary translations we scrambled through.

Next to Latin grammar the most painful subject I learnt was geography, which ought to have been the most interesting. It consisted almost entirely in learning by heart the names of the chief towns, rivers, and mountains of the various countries from, I think, Pinnock's "School Geography," which gave the minimum of useful or interesting information. It was something like learning the multiplication table both in the painfulness of the process and the permanence of the results. The incessant grinding in both, week after week and year after year, resulted in my knowing both the product of any two numbers up to twelve, and the chief towns of any English county so thoroughly, that the result was automatic, and the name of Staffordshire brought into my mind Stafford, Litchfield, Leek, as surely and rapidly as eight times seven brought fifty-six. The labour and mental effort to one who like myself had little verbal memory was very painful, and though the result has been a somewhat useful acquisition during life, I cannot but think that the same amount of mental exertion wisely directed might have produced far greater and more generally useful results.

History was very little better, being largely a matter of learning by heart names and dates, and reading the very baldest account of the doings of kings and queens, of wars, rebellions, and conquests. Whatever little knowledge of history I have ever acquired has been derived more from Shakespeare's plays and from good historical novels than from anything I learnt at school.

At one period when the family was temporarily broken

up, for some reason I do not remember, I was for about half a year a boarder in Mr. Crutwell's house, in company with twenty or thirty other boys; and I will here give the routine of a moderately good boarding-school at that period.

Our breakfast at eight consisted of a mug of milk-and-water and a large and very thick slice of bread-and-butter. For the average boy this was as much as he could eat, a few could not eat so much, a few wanted more, and the former often gave their surplus to the latter. Any boy could have an egg or a slice of bacon cooked if he bought it himself or had it sent from home, but comparatively very few had such luxuries.

For dinner at one o'clock we had hot joints of meat and vegetables for five days and hot meat-pies on Saturdays. On Sundays we had a cold joint of meat, with hot fruit-pies in the summer and plum-pudding in the winter, with usually some extra delicacy as custard or a salad. Every boy had half a pint of fairly good beer to drink, and any one who wished could have a second helping of meat.

At half-past five, I think, we had milk-and-water and bread-and-butter as at breakfast, from seven to eight we prepared lessons for the next day, and at eight o'clock we had supper, consisting of bread-and-cheese and, I think, another mug of beer.

Our regular games were cricket, baseball, leapfrog, high and long jumps, and, in the winter, turnpikes with hoops.

This latter was a means of enabling those who had no hoops to get the use of them. They kept turnpikes, formed by two bricks or stones placed the width of the foot apart, and the hoop-driver had to pass through without touching. If the hoop touched he gave it up, and kept the turnpike in his place. When there were turnpikes every five or ten yards all round the playground and a dozen or more hoops following each other pretty closely, the game was not devoid of its little excitements. We never played football (so far as I remember), which at that time was by no means such a common game as it is now. Among the smaller amusements which were always much liked were marbles and pegtops. The individuality of tops was rather curious, as some could only be made to spin by holding them with the peg upwards, others with it downwards, while others would spin when held in either position, and thrown almost anyhow. When tops were in fashion they might have been made the vehicle for very interesting teaching of mechanics, but that was quite beyond the range of the ordinary schoolmaster of the early part of the nineteenth century.

During my last year's residence at Hertford an arrangement was made by which, I suppose, the fees paid for my schooling were remitted on condition that I assisted in the school. I was a good writer and reader, and while continuing my regular classes in Latin and algebra, I took the younger boys in reading and dictation, arithmetic and writing. Although I had no objection whatever to the work itself, the anomalous position it gave me in the school—there being a score of boys older than myself who were scholars only—was exceedingly distasteful.

Another thing hurt me dreadfully at the time, because it exposed me to what I thought was the ridicule or contempt of the whole school. Like most other boys, I was reckless about my clothes, leaning my elbows on the desk till a hole was worn in my jacket, and, worse still, when cleaning my slate using my cuff to rub it dry. Slate sponges attached by a string were unknown to our school in those days. As new clothes were too costly to be had very often, my mother determined to save a jacket just taken for school wear by making covers for the sleeves, which I was to wear in school. These were made of black calico, reaching from the cuff to the elbow, and though I protested that I could not wear them, that I should be looked upon as a guy, and other equally valid reasons, they were one day put in my pocket, and I was told to put them on just before I entered the school. Of course I could *not* do it; so I brought them back and told my mother. Then after another day or two of trial, one morning the dreaded thunderbolt fell upon me. On entering school I was called up to the master's desk, he produced the dreaded calico sleeves, and told me that my mother wished me to wear them to save my jacket, and told me to put them on. Of course I had to do so. They fitted very well, and felt quite comfortable, and I dare say did not look so very strange. I have no doubt also that most of the boys had a fellow-feeling for me, and thought it a shame to thus make me an exception to all the school. But to me it seemed a cruel disgrace, and I was miserable so long as I wore them. How long that was I cannot remember, but while it lasted it was, perhaps, the severest punishment I ever endured.

In an article on the civilizations of China and Japan in *The Independent Review* (April, 1904), it is pointed out that the universal practice of "saving the face" of any kind of opponent rests upon the fundamental idea of the right of every individual to be treated with personal respect. With them this principle is taught from childhood, and pervades every class of society, while with us it is only recognized by the higher classes, and by them is rarely extended to inferiors or to children. The feeling that demands this recognition is certainly strong in many children, and those who have suffered under the failure of their elders to respect it, can well appreciate the agony of shame endured by the more civilized Eastern peoples, whose feelings are so often outraged by the total absence of all respect shown them by their European masters or conquerors. In thus recognizing the sanctity of this deepest of human feelings these people manifest a truer phase of civilization than we have attained to. Even savages often surpass us in this respect. They will often refuse to enter an empty house during the absence of the owner, even though something belonging to themselves may have been left in it; and when asked to call one of their sleeping companions to start on a journey, they will be careful not to touch him, and will positively refuse to shake him rudely, as an Englishman would have no scruple in doing.

As the period from the age of six to fourteen which I spent at Hertford was that of my whole home-life till I had a home of my own twenty-eight years later, and because it was in many ways more educational than the time I spent at school, I think it well to give a short account of it.

During the year or two spent at the first house we occupied in St. Andrew's Street very little occurred to impress itself upon my memory, partly, I think, because I was too young and had several playfellows of my own age, and partly, perhaps, because the very small house and yard at the back offered few facilities for home amusements. There was also at that time too much inequality between myself and my brother John for us to become such constant companions as we were a little later.

When we moved to the house beyond the Old Cross, nearly opposite the lane leading to Hartham, the conditions were altogether more favourable. The house itself was a more commodious one, and besides a yard at one side, it had a small garden at the back with a flower border at each side, where I first became acquainted with some of our common garden flowers. The gable end of the house in the yard, facing nearly south, had few windows, and was covered over with an old vine which not only produced abundance of grapes, but enabled my father to make some gallons of wine from the thinnings. But the most interesting feature of the premises to us two boys was a small stable with a loft over it, which, not being used except to store garden-tools and odd lumber, we had practically to ourselves. The loft especially was most delightful to us. It was reached by steps formed by nailing battens across the upright framing of the stable, with a square opening in the floor above. It thus required a little practice to climb up and down easily and to get a safe landing at top, and doing this became so familiar to us that we ran up and down it as quickly as sailors run up the shrouds of a vessel. Then the loft itself, under the slooping roof, gloomy and nearly dark in the remote corners, was almost like a robbers' cave; while a door opening to the outside by which hay could be pitched up out of a cart, afforded us plenty of light when we required it, together with the novel sensation and spice of danger afforded by an opening down to the floor, yet eight or nine feet above the ground.

This place was our greatest delight, and almost all the hours of daylight we could spare from school and meals were spent in it. Here we accumulated all kinds of odds and ends that might be useful for our various games or occupations, and here we were able to hide many forbidden treasures such as gunpowder, with which we used to make wild-fires as well as more elaborate fireworks. John was of a more mechanical turn than myself, and he used to excel in making all the little toys and playthings in which boys then used to delight. I, of course, looked on admiringly, and helped him in any way I could. I also tried to imitate him, but only succeeded in some of the simpler operations. Our most valuable guide was the "Boy's Own Book," which told us how to make numbers of things boys never think of making now, partly because everything is made for them, and also because children get so many presents of elaborate or highly ornamented toys when very young, that by the time they are old enough to make anything for themselves they are quite *blasé*, and can only be satisfied by still more elaborate and expensive playthings.

After my brother John went to London, and I was left alone at home, my younger brother being still too young for a playmate, I gave up most of these occupations, and began to develop a taste for reading. I still had one or two favourite companions with whom I used to go for long walks in the country, amusing ourselves in gravel or chalk pits, jumping over streams, and cutting fantastic walking-sticks out of the woods; but nothing afterwards seemed to make up for the quiet hours spent with my brother in the delightful privacy of the loft.

It was during our residence at this house near the Old Cross that, I think, my father enjoyed his life more than anywhere else at Hertford. Not only had he a small piece of garden and the fine grape-vine already mentioned, but there was a roomy brew-house with a large copper, which enabled him to brew a barrel of beer as well as make elder wine and grape-wine, bottle gooseberries, and other such work as he took great pleasure in doing. When here also, I think, he hired a small garden about half a mile off, where he could grow vegetables and small fruit, and where he spent a few hours of every fine day. And these various occupations were an additional source of interest and instruction to us boys. It was here, however, that our elder sister died of consumption in the year 1832, a little before she attained her twenty-second year. This was a severe loss to my father and mother, though I was not of an age to feel it much.

In the year 1834, a misfortune occurred that still further reduced the family income. Mr. Wilson, who had married my mother's only sister, was one of the executors of her father's will, and as he was a lawyer (the other executor being a clergyman), and his own wife and her sister were the only legatees, he naturally had the sole management of the property. Owing to a series of events with which we were only very imperfectly acquainted, he became bankrupt in this year, and his own wife and large family were at once reduced from a condition of comfort and even affluence to poverty almost as great as our own. But we children also suffered, for legacies of £100 each to my father's family, to be paid to us as we came of age, together with a considerable sum that had reverted to my mother on the death of her stepmother in 1828, had remained in Mr. Wilson's hands as trustee, and were all involved in the bankruptcy. He did all he possibly could for us, and ultimately, I believe, repaid a considerable part of the money, but while the legal proceedings were in progress, and they lasted fully three years, it was necessary for us to reduce expenses as much as possible. We had to leave our comfortable house and garden, and for a time had the use of half a rambling old house near All Saints' Church.

Before this, I think, my brother John had gone to London to be apprenticed, and the family at home consisted only of myself and my younger brother Herbert till my sister returned from France, where she had gone to study the language. It must have been about this time that I was sent for a few months as a boarder at the Grammar School, as already stated; but this whole period of my life is very indistinct. I am sure, however, that we moved to the next house in St. Andrew's Street early in 1836, because on May 15 of that year an annular eclipse of the sun occurred, visible in England, and I well remember the whole family coming out with smoked glasses into the narrow yard at the side of the house in order to see it. I was rather disappointed, as it only produced a peculiar gloom such as often occurs before a thunderstorm.

At midsummer, I think, we again moved to a part of a house next to St. Andrew's Church, where we had the

Silk family for neighbours in the larger half of the house. They also had most of the garden, on the lawn of which was a fine old mulberry tree, which in the late summer was so laden with fruit that the ground was covered beneath it, and I and my friend George used to climb up into the tree, where we could gather the largest and ripest fruit and feast luxuriously.

This was the last house we occupied in Hertford, the family moving to Hoddesdon some time in 1837, to a pretty but very small red-brick house called Rawdon Cottage, while I went to London and stayed at Mr. Webster's with my brother John, preparatory to going with my eldest brother William to learn land-surveying.

During the time I lived at Hertford I was subject to influences which did more for my real education than the mere verbal training I received at school. My father belonged to a book club, through which we had a constant stream of interesting books, many of which he used to read aloud in the evening. Among these I remember Mungo Park's travels and those of Denham and Clapperton in West Africa. We also had *Hood's Comic Annual* for successive years, and I well remember my delight with "The Pugsley Papers" and "A Tale of the Great Plague," while, as we lived first at a No. 1, I associated Hood's "Number One "with our house, and learnt the verses by heart when I was about seven years old. Ever since those early experiences I have been an admirer of

Hood in all his various moods, from his inimitable mixture of pun and pathos in his "Sea Spell," to the exquisite poetry of "The Haunted House," "The Elm Tree," and "The Bridge of Sighs."

We also had some good old standard works in the house, "Fairy Tales," "Gulliver's Travels," "Robinson Crusoe," and the "Pilgrim's Progress," all of which I read over again and again with constant pleasure. We also had the "Lady of the Lake," "The Vicar of Wakefield," and some others; and among the books from the club I well remember my father reading to us Defoe's wonderful "History of the Great Plague."

I think it was soon after we went to the Old Cross house that my father became librarian to a fairly good proprietary town library, to which he went for three or four hours every afternoon to give out and receive books and keep everything in order. After my brother John left home and I lost my chief playmate and instructor, this library was a great resource for me, as it contained a large collection of all the standard novels of the day. Every wet Saturday afternoon I spent there; and on Tuesdays and Thursdays, which were our four-o'clock days, I usually spent an hour there instead of stopping to play or going straight home. Sometimes I helped my father a little in arranging or getting down books, but I had most of the time for reading, squatting down on the floor in a corner, where I was quite out of the way. It was here that I read all Fenimore Cooper's novels, a great many of James's, and Harrison Ainsworth's "Rookwood," that fine highwayman's story containing a vivid account of Dick Turpin's Ride to York. It was here, too, I read the earlier stories of Marryat and Bulwer, Godwin's "Caleb Williams," Warren's "Diary of a Physician," and such older works as "Don Quixote," Smollett's "Roderick Random," "Peregrine Pickle," and "Humphry Clinker," Fielding's "Tom Jones," and Miss Burney's "Evelina." I also read, partially or completely, Milton's "Paradise Lost," Pope's "Iliad," Spenser's "Faërie Queene," and Dante's "Inferno," a good deal of Byron and Scott, some of the *Spectator* and *Rambler*, Southey's "Curse of Kehama," and, in fact, almost any book that I heard spoken of as celebrated or interesting. Walton's "Angler" was a favourite of my father's, and I well remember a woodcut illustration of Dove Dale with greatly exaggerated rocks and pinnacles, which made me long to see such a strange and picturesque spot—a longing which I only gratified about a dozen years ago, finding it more exquisitely beautiful than I had imagined it to be, even if not quite so fantastic.

My father and mother were old-fashioned religious people belonging to the Church of England, and, as a rule, we all went to church twice on Sundays, usually in the morning and evening. We also had to learn a collect every Sunday morning, and were periodically examined in our Catechism. On very wet evenings my father read us a chapter from the Bible and a sermon instead of the usual service.

Among our friends, however, were some Dissenters, and a good many Quakers, who were very numerous in Hertford; and on rare occasions we were taken to one of their chapels instead of to church, and the variety alone made this quite a treat. We were generally advised when some "friend" was expected to speak, and it was on such occasions that we visited the Friends' Meeting House, though I remember one occasion when, during the whole time of the meeting, there was complete silence. And when any brother or sister *was* "moved to speak," it was usually very dull and wearisome; and after having attended two or three times, and witnessed the novelty of the men and women sitting on opposite sides of the room, and there being no pulpit and no clergyman and no singing, we did not care to go again. But the Dissenters' chapel was always a welcome change, and we went there not unfrequently to the evening service. The extempore prayers, the frequent singing, and the usually more vigorous and exciting style of preaching was to me far preferable to the monotony of the Church service; and it was there only that, at one period of my life, I felt something of religious fervour, derived chiefly from the more picturesque and impassioned of the hymns. As, however, there was no sufficient basis of intelligible fact or connected reasoning to satisfy my intellect, this feeling soon left me, and has never returned.

CHAPTER IV

LONDON WORKERS, SECULARISTS AND OWENITES

HAVING finally left school at Christmas, 1836 (before I had completed my fourteenth year), I think it was early in 1837 that I was sent to London to live at Mr. Webster's in Robert Street, Hampstead Road, where my brother John was apprenticed. My father and mother were then about to move to the small cottage at Hoddesdon, and it was convenient for me to be out of the way till my brother William could arrange to have me with him to learn land-surveying.

Mr. Webster was a small master builder, who had a work-shop in a yard about five minutes' walk from the house, where he constantly employed eight or ten men preparing all the joinery work for the houses he built. At that time there were no great steam-factories for making doors and windows, working mouldings, etc., everything being done by hand, except in the case of the large builders and contractors, who had planing and sawing-mills of their own. Here in the yard was a sawpit in which two men, the top- and bottom-sawyers, were always at work cutting up imported balks of timber into the sizes required, while another oldish man was at work day after day planing up floor-boards. In the shop itself windows and doors, cupboards, staircases, and other joiner's work was always going on and the men employed all lived in the small streets surrounding the shop. The working hours were from six to half-past five, with one and a half hours out for meals, leaving a working day of ten hours.

Having nothing else to do, I used to spend the greater part of my time in the shop, seeing the men work, doing little jobs occasionally, and listening to their conversation. These were no doubt an average sample of London mechanics, and they were on the whole quite as respectable a set of men as any in a similar position to-day. I soon became quite at home in the shop, and got to know the peculiarities of each of the men. I heard their talk together, their jokes and chaff, their wishes and their ideas, and all those little touches of character which come out in the familiar intercourse of the workshop. My general impression is that there was very little swearing among them, much less than became common thirty years later, and perhaps about as much as among a similar class of men to-day. Neither was there much coarseness or indecency in their talk, far less indeed than I met with among professional young men a few years afterwards. One of the best of the workmen was a very loose character—a kind of Lothario or Don Juan by his own account—who would often talk about his adventures, and boast of them as the very essence of his life. He was a very good and amusing talker, and helped to make the time pass in the monotony of the shop; but occasionally, when he became too explicit or too boastful, the foreman, who was a rather serious though very agreeable man, would gently call him to order, and repudiate altogether his praises of the joys of immorality.

As my brother was, at the time I am now speaking of, nearly nineteen and a very good workman, he had complete liberty in the evenings after seven o'clock, the only limitation being that he had to be back about ten; while on special occasions he was allowed to take the door-key. He often took me with him on fine evenings to some of the best business streets in London to enjoy the shops, and especially to see anything of particular interest exhibited in them. But our evenings were most frequently spent at what was then termed a "Hall of Science," situated in John Street, Tottenham Court Road (now altered to Whitfield Street). It was really a kind of club or mechanics' institute for advanced thinkers among workmen, and especially for the followers of Robert Owen, the founder of the Socialist movement in England. Here we sometimes heard lectures on Owen's doctrines, or on the principles of secularism or agnosticism, as it is now called; at other times we read papers or books, or played draughts, dominoes, or bagatelle, and coffee was also supplied to any who wished for it. It was here that I first made acquaintance with some of Owen's writings, and especially with the wonderful and beneficent work he had carried on for many years at New Lanark. I also received my first knowledge of the arguments of sceptics, and read among other books Paine's "Age of Reason."

It must have been in one of the books or papers I read here that I met with what I dare say is a very old dilemma as to the origin of evil. It runs thus: "Is God able to prevent evil but not willing? Then he is not benevolent. Is he willing but not able? Then he is not omnipotent. Is he both able and willing? Whence then is evil?" This struck me very much, and it seemed quite unanswerable, and when at home a year or two afterwards, I took the opportunity one day to repeat it to my father, rather expecting he would be very much shocked at my acquaintance with any such infidel literature. But he merely remarked that such problems were mysteries which the wisest cannot understand, and seemed disinclined to any discussion of the subject. This, of course, did not satisfy me, and if the argument did not really touch the question of the existence of God, it did seem to prove that the orthodox ideas as to His nature and powers cannot be accepted.

I was also greatly impressed by a tract on "Consistency," written by Robert Dale Owen, the eldest son of Robert Owen, and as a writer superior in style and ability to his father. The chief object of it was to exhibit the horrible doctrine of eternal punishment as then commonly taught from thousands of pulpits by both the Church of England and Dissenters, and to argue that if those who taught and those who accepted such dogmas thoroughly believed them and realized their horror, all worldly pleasures and occupations would give way to the continual and strenuous effort to escape such a fate. I thoroughly agreed with Mr. Dale Owen's conclusion, that the orthodox religion of the day was degrading and hideous, and that the only true and wholly beneficial religion was that which inculcated the service of humanity, and whose only dogma was the brotherhood of man. Thus was laid the foundation of my religious scepticism.

Similarly, my introduction to advanced political views, founded on the philosophy of human nature, was due to the writings and teachings of Robert Owen and some of his disciples. His great fundamental principle, on which all his teaching and all his practice were founded, was that the character of every individual is formed *for* and not *by* himself, first by heredity, which gives him his natural disposition with all its powers and tendencies, its good and bad qualities; and, secondly, by environment, including education and surroundings from earliest infancy, which always modifies the original character for better or for worse. Of course, this was a theory of pure determinism, and was wholly opposed to the ordinary views, both of religious teachers and of governments, that, whatever the natural character, whatever the environment during childhood and youth, whatever the direct teaching, all men *could* be good if they liked, all *could* act virtuously, all *could* obey the laws, and if they wilfully transgressed any of these laws or customs of their rulers and teachers, the only way to deal with them was to punish them, again and again, under the idea that they could thus be *deterred* from future transgression. The utter failure of this doctrine, which has been followed in practice during the whole period of human history, seems to have produced hardly any effect on our systems of criminal law or of general education; and though other writers have exposed the error, and are still exposing it, yet no one saw so clearly as Owen how to put his views into practice; no one, perhaps, in private life has ever had such opportunities of carrying out his principles; no one has ever shown so much ingenuity, so much insight into character, so much organizing power; and no one has ever produced such striking results in the face of enormous difficulties as he produced during the twenty-six years of his management of New Lanark.

Of course, it was objected that Owen's principles were erroneous and immoral because they wholly denied free-will, because he advocated the abolition of rewards and punishments as both unjust and unnecessary, and because (it was argued) to act on such a system would lead to a pandemonium of vice and crime. The reply to this is that, acting on the principle of absolute free-will, every government has alike failed to abolish, or even to any considerable degree to diminish, discontent, misery, disease, vice, and crime; and that, on the other hand, Owen *did*, by acting on his own principle of the formation of character, transform a discontented, unhealthy, vicious, and wholly antagonistic population of 2500 persons to an enthusiastically favourable, contented, happy, healthy, and comparatively moral community, without ever having recourse to any legal punishment whatever, and without, so far as appears, discharging any individual for robbery, idleness, or neglect of duty; and all this was effected while increasing the efficiency of the whole manufacturing establishment, paying a liberal interest on the capital invested, and even producing a large annual surplus of profits which, in the four years 1809–13, averaged £40,000 a year, and only in the succeeding period, when the new shareholders agreed to limit their interest to 5 per cent, per annum, was this surplus devoted to education and the general well-being of the community.

Although most people have heard of New Lanark, few have any idea of Owen's work there or of the means by which he gradually overcame opposition and achieved the most remarkable results. It will, therefore, not be out of place to give a short account of his methods as explained in his autobiography, which I only had the opportunity of reading a few years ago.

In the year 1800, he became partner and sole manager of the New Lanark cotton-mills, and married the daughter of Mr. Dale, the former proprietor. Gradually, for many years, he had been elaborating his theory of human nature, and longing for an opportunity of putting his ideas in practice. And now he had got his opportunity. He had an extensive factory and workshops, with a village of about two thousand inhabitants all employed in the works, which, with about two hundred acres of surrounding land, belonged to the company. The character of the workers at New Lanark is thus described by Mr. W. L. Sargant, in his work "Robert Owen and his Social Philosophy," when describing the establishment of the mills about fifteeen years before Owen acquired them: "To obtain a supply of adult labourers a village was built round the works, and the houses were let at a low rent; but the business was so unpopular that few, except the bad, the unemployed, and the destitute, would settle there. Even of such ragged labourers the numbers were insufficient; and these, when they had learned their trade and become valuable, were self-willed and insubordinate." Besides these, there were about five hundred children, chiefly obtained from the workhouses of Edinburgh and other large towns, who were apprenticed for seven years from the age of six to eight, and were lodged and boarded in a large building erected for the purpose by the former owner, Mr. Dale, and were well attended to. But these poor children had to work from six in the morning to seven in the evening (with an hour and three-quarters for meals); and it was only after this task was over that instruction began. The children hated their slavery; many absconded; some were stunted, and even dwarfed in stature; and when their apprenticeship expired at the ages of thirteen to fifteen, they commonly went off to Glasgow or Edinburgh, with no natural guardians, and serving to swell the mass of vice and misery in the towns. "The condition of the families who had immigrated to the village was also very lamentable. The people lived almost without control in habits of vice, idleness, poverty, debt, and destitution. Some were drunk for weeks together. Thieving was general, and went on to a ruinous extent.... There was also a considerable drawback to the comfort of the people in the high price and bad quality of the commodities supplied in the village."

When Owen told his intimate friends who knew all these facts that he hoped to reform these people by a system of justice and kindness, and gradually to discontinue all punishment, they naturally laughed at him for a wild enthusiast; yet he ultimately succeeded to such an extent that hardly any one credited the accounts of it without personal inspection, and its fame spread over the whole civilized world. He had, besides the conditions already stated, two other great difficulties to overcome. The whole of the workers and overseers were strongly antagonistic to him as being an Englishman, whose speech they could hardly understand, and who, they believed, was sent to get more money for the owners and more work out of themselves.

They, therefore, opposed all he did by every means that ingenuity could devise, and though he soon introduced more order and regularity in the work and improved the quality of the yarn produced, they saw in all this nothing but the acts of a tool of the mill-owners somewhat cleverer, and therefore more to be dreaded, than those who had preceded him. An equally fierce opposition was made to any improvement in the condition of the houses and streets as to dirt, ventilation, drainage, etc. He vainly tried to assure the more intelligent of the overseers and workmen that his object was to improve their condition, to make them more healthy and happier and better off than they were. This was incredible to them, and for two years he made very little progress.

One thing, however, he did for the benefit of the workers which had some effect in disarming their enmity and suspicions. Instead of the retail shops where inferior articles were sold at credit for very high prices, he established stores and shops where every article of daily consumption was supplied at wholesale prices, adding only the cost of management. The result was that by paying ready money the people got far better quality at full 25 per cent, less than before; and the result soon became visible in their superior dress, improved health, and in the general comfort of their houses.

But what at length satisfied them that their manager was really their friend was his conduct when a great temporary scarcity of cotton and its rapid rise in price caused most of the mills to be shut, and reduced the workers to the greatest distress. But though Owen shut up the mills he continued to pay every worker full wages for the whole of the four months during which the scarcity lasted, employing them in thoroughly cleaning the mills and machinery, repairing the houses, etc. This cost £7000, which he paid on his own responsibility; but it so completely gained the confidence of the people that he was afterwards able to carry out improvements without serious obstruction. Being wholly opposed to infant labour, he allowed all arrangements with the guardians to expire, built a number of better houses, and. thus obtained families of workers to take the place of the children; but difficulties with the partners arose, the property was sold to a fresh set of partners, Owen being still the largest shareholder and manager, and a few years later again sold to Owen and a few of his personal friends, who agreed to allow him to manage the property, and to expend all profits above 5 per cent. for the benefit of the workers. Among his co-shareholders were Jeremy Bentham, with Joseph Foster and William Allen, well-known Quakers. It may be here stated that the property was purchased of Mr. Dale for £60,000, and was sold to Owen and his friends in 1814 for £114,100. This great increase of value was due in part to the large profits made by cotton mills generally at this period, and partly to Owen's skilful management and judicious expenditure.

He was now at last able to carry out his plans for the education of the children, none of whom he would allow to enter the mills as workers till they were ten years old. He built handsome and roomy schools, playrooms and lecture-rooms for infants from two to six, and for the older children from six to ten years old; and he obtained the best masters for the latter. The infant schools were superintended by himself, and managed by teachers he himself selected for their manifest love of children. His instructions to them were "that they were on no account ever to beat any one of the children, or to threaten them in any manner in word or action, or to use abusive terms, but were always to speak to them with a pleasant countenance, and in a kind manner and tone of voice; that they should tell the infants and children that they must on all occasions do all they could to make their playfellows happy; and that the older ones, from five to six years of age, should take especial care of the younger ones, and should assist to teach them to make each other happy." And these instructions, he assures us, were strictly followed by the man and woman he chose as infant-schoolmaster and mistress.

No books were to be used; but the children "were to be taught the uses and nature or qualities of the common things around them, by familiar conversation when the children's curiosity was excited so as to induce them to ask questions respecting them." The schoolrooms were furnished with paintings of natural objects, and the children were also taught dancing, singing, and military evolutions, which they greatly enjoyed. The children were never kept at any one occupation or amusement till they were fatigued and were taken much into the open air and into the surrounding country, where they were taught something about every natural object. Here we see all the essential features of the educational systems of Pestalozzi and Frœbel, worked out by his own observations of child-nature from his own childhood onward, and put into practice on the first opportunity with a completeness and success that were most remarkable.

The effect of his system on the adult workers was hardly less remarkable. To stop the continued pilfering of bobbins and other small articles in the mills, he invented a system (unfortunately not explained) by which the many thousands of these articles which passed from hand to hand daily were so recorded automatically that the loss of one by any particular worker could be always detected. In this way robbery large or small, was always discovered, *but no one was ever punished for it.* The certainty of discovery, however, prevented its being attempted, and it very soon ceased altogether.

Equally novel and ingenious was his method of avoiding the necessity for punishment, or even for a word of censure, for the many petty offences or infractions of rules that are inevitable in every large establishment. Owen calls it "the

silent monitor," but the workers called it the "telegraph."
Each superintendent of a department had a character-book,
in which the daily conduct of-every worker was set down
by marks for each of the ordinary offences, neglect of work,
swearing, etc., which when summed up gave a result in
four degrees—bad, indifferent, good, excellent. For every
individual there was a small wooden, four-sided tally, the sides
being coloured black, blue, yellow, and white, corresponding
to the above degrees of conduct. This tally was fixed at each
one's work-place, with the indicative colour only visible, so
that as Owen or his representative passed down the shops
at any time during the day, he could note at a glance the
conduct of each one during the preceding day, and thus get
both a general and a detailed view of the behaviour of the
workers. If any thought they were unfairly treated they could
complain to him, but in hardly any cases did this happen.
He tells us, "As I passed through all the rooms, the workers
always observed me look at these telegraphs, and when black
I merely looked at the person, and then at the colour, but
never said a word to one of them by way of blame. At first,"
he says, "a large proportion daily were black and blue, few
yellow, and scarcely any white. Gradually the blacks were
changed for blue, the blues for yellow, and the yellows for
white. Soon after the adoption of this telegraph I could at
once see by the expression of countenance what was the
colour which was shown. As there were four colours there

were four different expressions of countenance, most evident to me as I passed along the rooms.

... Never perhaps in the history of the human race has so simple a device created in so short a period so much order, virtue, goodness, and happiness, out of so much ignorance, error, and misery. And for many years the permanent daily conduct of a very large majority of those who were employed deserved, and had, No. I placed as their character on the books of the company."

Every visitor to New Lanark who published any account of his observations seems to have agreed as to the exceptional health, good conduct, and well-being of the entire population; while residents in the vicinity, as well as the ruling authorities of the district, bore witness that vice and crime were almost wholly unknown. And it must be remembered that this was all effected upon the chance population found there, which was certainly no better if no worse than the usual lowest class of manufacturing operatives at that period. There appears to have been not a single case of an individual or a family being expelled for bad conduct; so that we are compelled to trace the marvellous improvement that occurred entirely to the *partial* application of Owen's principles of human nature, most patiently and skilfully applied by himself. They were necessarily only a partial application, because a large number of the adults had not received the education and training from infancy which was essential for producing their full beneficial results. Again, the whole establishment was a manufactory, the property of private capitalists, and the adult population suffered all the disadvantages of having to work for long hours at a monotonous employment and at low rates of wages, circumstances wholly antagonistic to any full and healthy and elevated existence. Owen used always to declare that the beneficial results at which all visitors were so much astonished were only one-tenth part of what *could* and *would* be produced if his principles were fully applied. If the labour of such a community, or of groups of such communities, had been directed with equal skill to produce primarily the necessaries and comforts of life for its own inhabitants, with a surplus of such goods as they could produce most economically, in order by their sale in the surrounding district to be able to supply themselves with such native or foreign products as they required, then each worker would have been able to enjoy the benefits of change of occupation, always having some alternation of outdoor as well as indoor work; the hours of labour might be greatly reduced, and all the refinements of life might have been procured and enjoyed by them.

The one great error Owen committed was giving up the New Lanark property and management, and spending his large fortune in the endeavour to found communities in various countries of chance assemblages of adults, which his own principles should have shown him were doomed to failure. He always maintained that a true system of education from infancy to manhood was *essential* to the best formation of character. His infant schools had only been about ten years in existence, when, owing to some difficulties with his Quaker partners, who had always objected to the dancing and drill, he gave up the management into their hands.

Notwithstanding this one fatal error, an error due to the sensitive nobility of his character and to his optimistic belief in the power of truth to make its way against all adverse forces, Robert Owen will be remembered as one of the wisest, noblest, and most practical of philanthropists, as well as one of the best and most lovable of men.

I have a recollection of having once heard him give a short address at this "Hall of Science," and that I was struck by his tall spare figure, very lofty head, and highly benevolent countenance and mode of speaking. Although later in life my very scanty knowledge of his work was not sufficient to prevent my adopting the individualist views of Herbert Spencer and of the political economists, I have always looked upon Owen as my first teacher in the philosophy of human nature and my first guide through the labyrinth of social science. He influenced my character more than I then knew, and now that I have read his life and most of his works, I am fully convinced that he was the greatest of social reformers and the real founder of modern Socialism. For these reasons

I trust that my readers will not consider the space I have here devoted to an outline of his great work at New Lanark is more than the subject deserves.

The preceding sketch of his work is founded upon his "Life" written by himself, and accompanied by such a mass of confirmatory reports and correspondence as to show that it can be thoroughly relied on. It has, however, long been out of print, and very few people have read it or even heard of it, and it is for this reason that I have given this brief outline.

CHAPTER V

SURVEYING IN BEDFORDSHIRE

IT was, I think, early in the summer of 1837 that I went with my brother William into Bedfordshire to begin my education as a land-surveyor. The first work we had was to survey the parish of Higham Gobion for the commutation of the tithes. It was a small parish of about a thousand acres, with the church, vicarage, and a good farmhouse on the highest ground, and a few labourers' cottages scattered about, but nothing that could be called a village. The whole parish was one large farm; the land was almost all arable and the fields very large, so that it was a simple piece of work. We took up our quarters at the Coach and Horses public-house in the village of Barton-in-the-Clay, six miles north of Luton, on the coach-road to Bedford. We were nearly a mile from the nearest part of the parish, but it was the most convenient place we could get.

An intelligent young labourer was hired to draw the chain in measuring, while I carried a flag or measuring-rod and stuck in pegs or cut triangular holes in the grass, where required, to form marks for future reference. We carried bill-hooks for cutting rods and pegs, as well as for clearing away branches that obstructed the view, and for cutting gaps in the hedges on the main lines of the survey, in order to lay them out perfectly straight. We started work after an early breakfast, and usually took with us a good supply of bread-and-cheese and half a gallon of beer, and about one o'clock sat down under the shelter of a hedge to enjoy our lunch. My brother was a great smoker, and always had his pipe after lunch (and often before breakfast), and, of course, the chain-bearer smoked too. It therefore occurred to me that I might as well learn the art, and for some days I tried a few whiffs. Then, going a little too far, I had such a violent attack of headache and vomiting that I was cured once and for ever from any desire to smoke, and although I afterwards lived for some years among Portuguese and Dutch, almost all of whom are smokers, I never felt any inclination to try again.

It was while living at Barton that I obtained my first information that there was such a science as geology, and that chalk was not *everywhere* found under the surface, as I had hitherto supposed. My brother, like most land-surveyors, was something of a geologist, and he showed me the fossil oysters of the genus Gryphæa and the Belemnites, which we had hitherto called "thunderbolts," and several other fossils which were abundant in the chalk and gravel around Barton. While here I acquired the rudiments of surveying and mapping, as well as calculating areas on the map by the rules of trigonometry. This I found very interesting work, and it was rendered more so by a large volume belonging to my brother giving an account of the great Trigonometrical Survey of England, with all the angles and the calculated lengths of the sides of the triangles formed by the different stations on hilltops, and by the various church spires and other conspicuous objects. The church spires of Barton and Higham Gobion had been thus used, and the distance between them accurately given; and as the line from one to the other ran diagonally across the middle of the parish we were surveying, this was made our chief base-line, and the distance as measured found to agree very closely with that given in the survey. This volume was eagerly read by me, as it gave an account of all the instruments used, including the great theodolite three feet in diameter for measuring the angles of the larger triangles formed by distant mountain tops often twenty or thirty miles apart, and in a few cases more than a hundred miles; the accurate measurement of the base-lines by steel chains laid in wooden troughs, and carefully tightened by exactly the same weight passing over a pulley, while the ends were adjusted by means of microscopes; the exact temperature being also taken by several thermometers in order to allow for contraction or expansion of the chains; and by all these refinements several base-lines of seven or eight miles in length were measured with extreme accuracy in distant parts of the country. These base-lines were tested by repeated measurements in opposite directions, which were found to differ only by about an inch, so that the mean of all the measurements was probably correct to less than half that amount.

These bases were connected by the system of triangulation already referred to, the angles at all the stations being taken with the best available instruments and often repeated by different observers, while allowance had also to be made for height above the sea-level, to which all the distances had to be reduced. In this way, starting from any one base, the lengths of the sides of all the triangles were calculated, and ultimately the length of the other bases; and if there had been absolutely no error in any of the measurements of base-lines or of angles, the length of a base obtained by calculation would be the same as that by direct measurement. The results obtained showed a quite marvellous accuracy. Starting from the base measured on Salisbury Plain, the length of another base on the shore of Lough Foyle in the north of Ireland was calculated through the whole series of triangles connecting them, and this calculated length was found to differ from the measured length by only five inches and a fraction. The distance between these two base-lines is about three hundred and sixty miles.

It was here, too, that during my solitary rambles I first began to feel the influence of nature and to wish to know more of the various flowers, shrubs, and trees I daily met with, but of which for the most part I did not even know the English names. At that time I hardly realized that there was such a science as systematic botany, that every flower and every meanest and most insignificant weed had been accurately described and classified, and that there was any kind of system or order in the endless variety of plants and animals which I knew existed.

Barton was a rather large straggling village of the old-fashioned, self-contained type, with a variety of small tradesmen and mechanics, many of whom lived in their own freehold or leasehold houses with fair-sized gardens. Our landlord was a Radical, and took a newspaper called *The Constitutional*, which was published at Birmingham, and contained a great deal of very interesting matter. This was about the time the dean and chapter refused to allow a monument to be erected to Byron in Westminster Abbey, which excited much indignation among his admirers. One of these wrote some lines on the subject which struck me as being so worthy of the occasion that I learnt them by heart, and by constant repetition (on sleepless nights) have never forgotten them. They were printed in the newspaper without a signature, and I have never been able to learn who was the author of them. I give them here to show the kind of poetry I admired then and still enjoy—

"Away with epitaph and sculptured bust!
Leave these to decorate the mouldering dust

Of him who needs such substitutes for fame—
The chisel's pomp to deck a worthless name.
Away with these! A Byron needs them not;
Nature herself selects a deathless spot,
A nation's heart: the Poet cannot die,
His epitaph is Immortality.
What are earth's mansions to a tomb like this?
When time hath swept into forgetfulness
Wealth-blazoned halls and gorgeous cemeteries,
The mouldering Abbey with its sculptured lies,
His name, emblazoned in the wild, the free,
The deep, the beautiful of earth, shall be
A household word with millions. Dark and wild
His song at times, his spirit was the child
Of burning passion. Yet when he awoke
From his dark hours of bondage, when he broke
His cage and seized his harp, did he not make
A peal of matchless melody and shake
The very earth with joy? Still thrills the heart
Of man at those sweet notes; sacred despots start
To curse them from their thrones; they pierce the cell
And cheer the captive in his chains; they tell
Lessons of life to struggling liberty.
Death mars the man but spares his memory,
Nor tears one laurel from his wreath of fame.
How many glorious thoughts of his we claim

Our heritage for ever; beacon lights
To guide the barque of freedom through the nights
Of tyranny and woe, when not a star
Of hope looks down to glad the mariner:
Thoughts which must ever haunt us, like some dream
Of childhood which we ne'er forget, a gleam
Of sunshine flashing o'er life's troubled stream!"

Those who only know Byron by his more romantic or pathetic poems, and who may think the panegyric of the anonymous writer in *The Constitutional** to be overdrawn, should read "The Age of Bronze," which is pervaded throughout with the detestation of war, with admiration of those who fought only for freedom, and with scorn and contempt for the majority of English landlords, who subordinated all ideas of justice or humanity to the keeping up of their rents. Even if it stood alone, this one poem would justify the poet as an upholder of the rights of man and as a truly ethical teacher.

As there was no work of importance after the maps and reference books of the parish we had been surveying were completed and delivered, and winter was approaching, I went home for a short holiday. My father and mother and my younger brother were then living in Hoddesdon, and as there was no direct conveyance I made the journey on foot. It was, I think, the end of November, and as the distance was about thirty miles, and I was not very strong, I took two days, sleeping on the way at a roadside public-house. I went through Hitchin and Stevenage, and near the former place passed a quarry of a reddish chalk almost as hard as marble, which was used for building. This surprised me, as I had hitherto only seen the soft varieties of chalk, and had been accustomed to look upon it as more earth than stone. The only other thing that greatly interested me was a little beyond Stevenage, where, on a grassy strip by the roadside, were six ancient barrows or tumuli, which I carefully inspected; and whenever I have since travelled by the Great Northern Railway, I have looked out for these six tumuli, near which the line passes.

* This newspaper—*The Constitutional*—appears to have existed only two years. The *Daily News*, referring to a sale of Thackeray rarities last year, states that he contributed several articles to that paper as Paris correspondent, and that, in consequence, a set of the paper sold in 1899 for two hundred guineas. A friend informs me that it does not exist in the Bodleian Library.

After a few weeks at home at Hoddesdon, I went back to Barton, where we had some work till after Christmas. On New Year's Day, 1838, the first section of the London and Birmingham Railway was opened to Tring, and I and my brother took advantage of it to go up to London, where he had some business, and the next day I walked to Hoddesdon for a short holiday. My brother while in London obtained the survey for tithe commutation of a parish in Bedfordshire, where I was to meet him on the 14th or 15th of January, at the village of Turvey, eight miles beyond Bedford.

I had first to go back to Barton to pay a few bills and pack up the books, instruments, etc., we had left there to be sent by carrier's waggon. I therefore left home on the 12th, and I think walked back to Barton, and the next day did what was required, took leave of my friends there, and on the morning of the 14th, after an early breakfast, started to walk to Turvey through Bedford, a distance of about twenty miles. I dined at Bedford, and reached Turvey before dark.

For the next six days we were at work laying out the main lines for the survey of the parish, cutting hedges, ranging flags, ascertaining boundaries, and beginning the actual measurements.

A curious incident may here be noted. One day I was out on the frozen meadows across the river Ouse, assisting in marking out one of our main lines which had to cross the windings of the river, when I saw a pleasant-looking young man coming towards me carrying a double-barrelled gun. When he was a few yards off, two very large birds, looking like wild geese, came flying towards us, and as they passed overhead at a moderate height, he threw up his gun, fired both barrels, and brought them both to the ground. Of course I went up to look at them, and found they were a fine pair of wild swans, the male being about five feet long from beak to end of tail. "That was a good shot," I remarked; to which he replied, "Oh! you can't miss them, they are as big as a barn door." Afterwards I found that this was young Mr. H. H. Higgins, of Turvey Abbey, his father being one of the principal landowners in the parish.

More than half a century later (in November, 1889), I was invited to Liverpool to give some lectures, and some time before the date fixed upon I received a very kind letter from the Rev. H. H. Higgins, inviting me to dine with him on my arrival, and offering to assist me in every way he could. I declined the invitation, but told him what hotel I was going to, and said that I should be glad to see him. His letter recalled to me my acquaintance at Turvey, but I did not see how a Liverpool clergyman could have any close relationship to a wealthy Bedfordshire landowner. I found Mr. Higgins at the station with a carriage ready, and he told me that, as I did not wish to go out to dinner, he and some friends had taken the liberty of ordering a dinner at my hotel, and hoped I would dine with them. He was as pleasant as an old friend, and of course I accepted.

When his friends left about an hour after dinner, I asked him, if he had no engagement, to stay a little longer, as I wished to find out the mystery. I then asked him if he knew a place called Turvey, in Bedfordshire, to which he replied, "I ought to know it, for I was born there, and my father owned the estate there to which I am heir." I then felt pretty sure of my man, and asked him if he remembered, during a very hard frost about fifty years ago, shooting a pair of wild swans at Turvey. "Why, of course I do," said he. "But how do you know it?" "Because I was there at the time and saw you shoot them. Do not you remember a thin tall lad who came up to you and said, 'That was a good shot,' and you replied, 'Oh! you can't miss them, they are as big as a barn door'?" "No," he said, "I don't remember you at all, but that is just what I should have said." His delight was great, for his story of how he shot the two wild swans was not credited even by his own family, and he made me promise to go to his house after the lecture on the next night, and prove to them that he had not been romancing. When I went, I was duly introduced to his grown-up sons and daughters as one who had been present at the shooting of the swans, which I had been the first to mention. That was a proud moment for the Rev. H. H. Higgins, and a very pleasant one to myself.

Soon after we came to Turvey a young gentleman from Bedford came to us to learn a little surveying. He was, I think, the son of an auctioneer or estate agent, and was about eighteen or twenty years old. As my brother was occasionally away for several days at a time when we sometimes had nothing to go on with, he would amuse himself fishing, of which he was very fond. Sometimes I went with him, but I usually preferred walking about the country, though I cannot remember that I had at this time any special interest in doing so. He often caught some large coarse fish, such as bream or pike, which were the commonest fish in the river, but were hardly worth eating. Towards the latter part of our survey in the spring months, my brother left us a portion of the work to do by ourselves when he was away for a week or two. I was therefore left mostly to the companionship of our temporary pupil, and he, like the majority of the young men I met at this period of my life, was by no means an edifying acquaintance.

But, notwithstanding that I was continually thrown into such society from the time I left school, I do not think it produced the least bad effect upon my character or habits in after-life. This was partly owing to natural disposition, which was reflective and imaginative, but more perhaps to the quiet and order of my home, where I never heard a rude word or an offensive expression. The effect of this was intensified by my extreme shyness, which made it impossible for me to use words or discuss subjects which were altogether foreign to my home-life, as a result of which I have never been able to use an oath, although I have frequently felt those impulses and passions which in many people can only find adequate expression in such language. This, I think, is a rather striking example of the effects of home influence during childhood, and of that *kind* of education on which Robert Owen depended for the general improvement of character and habits.

It was some time in May or June of 1838 that we left Turvey for Silsoe, where my brother had some temporary work. This very small village is an appanage of Wrest Park, the seat of Earl de Grey, and is about halfway between Luton and Bedford. It consisted of a large inn with a considerable posting business, a few small houses, cottages, and one or two shops, and, like most such villages, it is no larger to-day than it was then.

Our work here was mainly copying maps or making surveys connected with the estate, and for this purpose we had the use of a small empty house nearly opposite the inn, where a large drawing-table and a few chairs and stools were all the furniture we required. Here we used sometimes to sit of a summer's evening with one or two friends for privacy and quiet conversation.

One day, having to drive over to Dunstable on some business, my brother took me with him. When there, we walked out to a deep cutting through the chalk about a mile to the north-west, where the road was being improved by further excavation to make the ascent easier. This was the great mail-coach road to Birmingham and Holyhead, and although the railway from London to Birmingham was then making and partly finished, nobody seemed to imagine that in twelve years more a railway would be opened the whole distance, and, so far as the mails and all through traffic were concerned, all such costly improvement on the high-roads would be quite unnecessary.

My brother had some conversation with the engineer who was inspecting the work, and took a lump of chalk home with him to ascertain its specific gravity, as to which there was some difference of opinion. While taking luncheon at the hotel we met a gentleman named Matthews about my brother's age, who turned out to be a surveyor, and who was also interested in engineering generally; and after luncheon they borrowed a small pair of scales and a large jug of water, and by suspending the chalk by a thread below the scale-pan, they weighed it in water, having first weighed it dry in the ordinary way, and the weight in air, divided by the difference between the weights in air and water, gives the specific gravity sufficiently near for ordinary purposes. This little experiment interested me greatly, and made me wish to know something about mechanics and physics. Mr. Matthews lived at Leighton Buzzard, where he carried on the business of watch and clock maker as well as that of engineer and surveyor. He had undertaken the survey of the parish of Soulbury, but having too much other work to attend to, he was looking out for some one to take it off his hands. This matter was soon agreed upon, and a few weeks afterwards we left Silsoe to begin the work.

The village of Soulbury is a very small one, though the parish is rather large. It is only three miles from Leighton, and we obtained accommodation in the school-house.

The district was rather an interesting one. The parish was crossed about its centre by the small river Ouzel, a tributary of the Ouse, bordered by flat verdant meadows, beyond which the ground rose on both sides into low hills, which to the north-east reached five hundred feet above the sea, and being of a sand formation, were covered with heaths and woods of fir trees. Parallel with the river was the Grand Junction Canal, which at that time carried all the heavy goods from the manufacturing districts of the Midlands to London. Following the same general direction, but about half a mile west on higher ground, the London and Birmingham Railway was in course of construction, a good deal of the earthwork being completed, most of the bridges built or building, and the whole country enlivened by the work going on.

At the same time the canal had been improved at great cost to enable it to carry the increased trade that had been caused by the rapid growth of London and the prosperity of agriculture during the early portion of the nineteenth century. About thirty miles further on the watershed between the river-basins of the Ouse and Severn had to be crossed, a district of small rainfall and scanty streams, from which the whole supply of the canal, both for its locks as well as for evaporation and leakage, had to be drawn. Whenever there was a deficiency of water here to float the barges and fill the locks, traffic was checked till the canal filled again; and this had become so serious that, for a considerable portion of the canal, it had been found necessary to erect steam-engines to pump up the water at every lock from the lower to the higher level. Sometimes there were two, three, or more locks close together, and in these cases a more powerful engine was erected to pump the water the greater height. Up to this time I had never seen a steam-engine, and therefore took the greatest interest in examining these both at rest and at work. They had been erected by the celebrated firm of Boulton and Watt, and were all of the low-pressure type then in use, with large cylinders, overhead beam, and parallel motion, but each one having its special features, the purport of which was explained to me by my brother, and gave me my first insight into some of the more important applications of the sciences of mechanics and physics.

Of course at that time nobody foresaw the rapid development of railways all over the country, or imagined that they could ever compete with canals in carrying heavy goods. Yet within two years after the completion of the line to Birmingham, the traffic of the canal had decreased to 1,000,000 tons, while it was 1,100,000 tons in 1837. Afterwards it began slowly to rise again, and had reached

1,627,000 tons in 1900, an exceedingly small increase as compared with that of the railway. And this increase is wholly due to local traffic between places adjacent to the canal.

In the northern part of the parish, which extended nearly to the village of Great Brickhill, were some curious dry valleys with flat bottoms, and sides clothed with fir woods, a kind of country I had not yet seen, and which impressed me as showing some connection between the geological formation of the country and its physical features, though it was many years later when, by reading Lyell's "Principles of Geology," I first understood *why* it should be so. Another interesting feature of the place, of which no one then saw the significance, was a large mass of hard conglomerate rock, or pudding-stone, which lay in the centre of the spot where the three roads met in front of the house where we lodged. It was roughly about a yard in diameter and about the same height, and had probably at some remote period determined the position of the village and the meeting-point of the three roads. Being a kind of rock quite different from any found in that part of England, it was probably associated with some legend in early time, but it is in all probability a relic of the Ice Age, and was brought by the glacier or ice-sheet that at one time extended over all midland England as far as the Thames valley. But at this time not a single British geologist knew anything about a glacial epoch, it being two years later, in 1840, when Louis Agassiz showed Dr. Buckland such striking indications of ice-action in Scotland as to convince him of the reality of such a development of glaciers in our own country at a very recent period.

Having finished our plans of Soulbury, and made the three copies needed with their books of reference, with some other odd work, my brother took me up to London on Christmas Eve, travelling by coach to Berkhampstead and thence on to London by the railway, which had been just opened. We went third class for economy, in open trucks identical with modern goods trucks, except that they had hinged doors, but with no seats whatever, so that any one tired of standing must sit upon the floor. Luckily it was mild weather, and the train did not go more than fifteen or twenty miles an hour, yet even at that pace the wind was very disagreeable. The next day we went home to Hoddesdon for a holiday. It had been settled that, as no more surveying work was in view, I should go back to Leighton to Mr. Matthews for a few months to see if I should like to learn the watch and clock making business as well as surveying and general engineering; and as there seemed to be nothing else available, I did so.

Mr. William Matthews was a man of about thirty. He had been married two years, and had a little girl under a year old. Both he and Mrs. Matthews were pleasant people, and I felt that I should be comfortable with them. Mr. Matthews had also charge of the town gas-works, which involved some knowledge of practical chemistry, and a good deal of mechanical work. I spent about nine months in his house, and during that time learnt to take an ordinary watch to pieces, clean it properly, and put it together again, and the same with a clock; to do small repairs to jewellery; and to make some attempts at engraving initials on silver. I also saw the general routine of gas manufacture; but hardly any surveying, which was the work I liked best. I was, therefore, very glad when circumstances, not connected with myself, put an end to the arrangement. Mr. Matthews accepted the offer of a partnership in an old-established wholesale watchmaking firm in the city of London, and gave up his Leighton business.

This may be considered the first of several turning-points of my life, at which, by circumstances beyond my own control, I have been insensibly directed into the course best adapted to develop my special mental and physical activities. If I had been apprenticed to Mr. Matthews I should have become a mechanical tradesman in a country town, by which my life would almost certainly have been shortened and my mental development stunted by the monotony of my occupation.

CHAPTER VI

RADNORSHIRE

IN the autumn of 1839 my brother came to Leighton to take me away, and in a day or two we started for Herefordshire, going by the recently opened railroad to Birmingham, where we visited an old friend of my brother's, a schoolmaster, whose name I forget, and who I remember showed us with some pride how his school was warmed by hot-water pipes, then somewhat unusual. We then went on by coach through Worcester to Kington, a small town of about two thousand inhabitants, only two miles from the boundary of Radnorshire. It is pleasantly situated in a hilly country, and has a small stream flowing through it. Just beyond the county boundary, on the road to Old and New Radnor, there is an isolated craggy hill called the Stanner Rocks, which, being a hard kind of basalt very good for road-metal, was being continually cut away for that purpose. It was covered with scrubby wood, and was the most picturesque object in the immediately surrounding country.

In a solitary letter, accidentally preserved, written at this time to my earliest friend, George Silk, I find the following passage which well expresses the pleasure I felt in getting back to land-surveying:—

" I think you would like land-surveying, about half indoors and half outdoors work. It is delightful on a fine summer's day to be (literally) 'cutting' all over the country, following the chain and admiring the beauties of nature, breathing the fresh and pure air on the hills, or in the noontide heat enjoying our luncheon of bread-and-cheese in a pleasant valley by the side of a rippling brook. Sometimes, indeed, it is not quite so pleasant on a cold winter's day to find yourself on the top of a bare hill, not a house within a mile, and the wind and sleet chilling you to the bone. But it is all made up for in the evening; and those who are in the house all day can have no idea of the pleasure there is in sitting down to a good dinner and feeling hungry enough to eat plates, dishes, and all."

Some time during the winter I went alone to correct an old map of the parish of New Radnor. This required no regular surveying, but only the insertion of any new roads, buildings, or divisions of fields, and taking out any that had been cleared away. As these changes were not numerous and the new fences were almost always straight lines, it was easy to mark on the map the two ends of such fences by measuring from the nearest fixed point with a ten or fifteen-link measuring-rod, and then drawing them in upon the plan. Sometimes the direction was checked by taking an angle with the pocket sextant at one or both ends, where one of these could not be seen from the other. As the whole plan was far too large to be taken into the field, tracings were made of portions about half a mile square, which were mounted on stiff paper or linen, and folded up in a loose cover for easy reference. In this way a whole parish of several thousand acres could be examined and corrected in a week or two, especially in a country like Wales, where, from a few elevated points, large tracts could be distinctly seen spread out below, and any difference from the old map be easily detected. I liked this kind of work very much, as I have always been partial to a certain amount of solitude, and am especially fond of rambling over a country new to me.

New Radnor, though formerly a town of some importance, was then, and I believe is still, a mere village, and a poor one, Presteign being the county town. It is situated on the southern border of Radnor Forest, a tract of bare mountains about twenty square miles in extent, the highest point being a little over two thousand feet above the sea. Over a good deal of this country I wandered for about a week, and enjoyed my work very much.

Early the next year, I think about February, my brother and I went to do some surveying at Rhaidr-Gwy (now commonly called Rhayader), a small town in Radnorshire on the Upper Wye, and only fifteen miles from its source in the Plynlymmon range. A young man from Carmarthenshire came to us here to learn surveying, and to him I probably owe my life. One day, I think on a Sunday afternoon, we were walking together up a rocky and boggy valley which extended some miles to the west of the town. As we were strolling along, picking our way among the rocks and bog, I inadvertently stepped upon one of those small bog eyeholes which abound in such places, and are very dangerous, being often deep enough to swallow up a man, or even a horse. One leg went in suddenly up to the hip, and I fell down, but fortunately with my other leg stretched out upon the surface. I was, however, in such a position that I could not rise, and had I been alone my efforts to extricate myself might easily have drawn my whole body into the bog, as I could feel no bottom to it. But my companion easily pulled me out, and we walked home, and thought little of it. It had, however, been a hard frost for some time, and the mud was ice-cold, and after a few days I developed a bad cough with loss of appetite and weakness. The local doctor was a friend of ours, and he gave me some medicine, but it did no good, and I got worse and worse, with no special pain, but with a disgust of food, and for more than a week I ate nothing but perhaps a small biscuit each day soaked in tea without milk, though always before and since I greatly disliked tea without milk. At length the doctor became frightened, and told my brother that he could do nothing for me, and that he could not be answerable for my life. He added that he knew but one man who could save me, a former teacher of his, Dr.

Ramage, who was the only man who could cure serious lung disease, though he was considered a quack by his fellow-practitioners.

As I got no better, a few days later we started for London, I think sleeping at Birmingham on the way. On going to Dr. Ramage, who tested my lungs, etc., he told my brother that he was just in time, for that in a week more he could probably not have saved me, as I had an extensive abscess of the lungs. His treatment was very simple but most effective, and was the forerunner of that rational treatment by which it is now known that most lung diseases are curable. He ordered me to go home to Hoddesdon immediately, to apply half a dozen leeches to my chest at a place he marked with ink, and to take a bitter medicine he prescribed to give me an appetite; but these were only preliminaries. The essential thing was the use of a small bone breathing-tube, which he told us where to buy, and which I was to use three times a day for as many minutes as I could without fatigue; that I was to eat and drink anything I fancied, be kept warm, but when the weather was mild sit out-of-doors. I was to come back to him in a week.

The effect of his treatment was immediate. I at once began to eat, and though I could not breathe through the tube for more than a minute at first, I was soon enabled to increase it to three and then to five minutes. It was constructed with a valve so that the air entered freely, but passed out slowly so that it was kept in the lungs for a few seconds at each inspiration. When I paid my second visit to Dr. Ramage, he told me that I was getting on well, and need not come to him again, that I was to continue using the breathing-tube for five minutes three or four times a day. He also strongly advised me, now I saw the effect of deep and regular breathing, to practise breathing in the same way without the tube, and especially to do so when at leisure, when lying down, or leaning back in an easy-chair, and to be sure to fill my lungs well and breathe out slowly. "The natural food of the lungs," he said, "is fresh air. If people knew this, and acted upon it, there would be no consumption, no lung disease." I have never forgotten this. I have practised it all my life (at intervals), and do so still, and I am sure that I owe my life to Dr. Ramage's treatment and advice.

In about two months I was well again, and went back to Kington, and after a little office-work my brother and I went to the little village of Llanbister, near the middle of Radnorshire, the nearest towns being Builth, in Brecknockshire, and Newtown in Montgomeryshire, both more than twelve miles distant. This was a very large parish, being fifteen miles long, but I think we could only have corrected the old map or we should have been longer there than we really were. Here, also, we had a young gentleman with us for a month or two to practise surveying. He was, I think, a Welshman, and a pleasant and tolerably respectable young man, but he had one dreadful habit—excessive smoking. I have never met a person so much a slave to the habit, and even if I had had any inclination to try it again after my first failure, his example would have cured me.

He prided himself on being a kind of champion smoker, and assured us that he had once, for a wager, smoked a good-sized china teapot full of tobacco through the spout.

When we had finished at Llanbister, we went about ten miles south to a piece of work that was new to me—the making of a survey and plans for the inclosure of common lands. This was at Llandrindod Wells, where there was then a large extent of moor and mountain surrounded by scattered cottages with their gardens and small fields, which, with their rights of common, enabled the occupants to keep a horse, cow, or a few sheep, and thus make a living. All this was now to be taken away from them, and the whole of this open land divided among the landowners of the parish or manor in proportion to the size or value of their estates. To those that had much, much was to be given, while from the poor their rights were taken away; for though nominally those that *owned* a little land had some compensation, it was so small as to be of no use to them in comparison with the grazing rights they before possessed. In the case of all cottagers who were tenants or leaseholders, it was simple robbery, as they had no compensation whatever, and were left wholly dependent on farmers for employment. And this was all done—as similar inclosures are almost always done—under false pretences. The "General Inclosure Act" states in its preamble, "Whereas it is expedient to facilitate the inclosure and improvement of commons and other lands now subject to the rights of property which obstruct cultivation and the productive employment of labour, be it enacted," etc. But in hundreds of cases, when the commons, heaths, and mountains have been partitioned out among the landowners, the land remains as little cultivated as before. It is either thrown into adjacent farms as rough pasture at a nominal rent, or is used for game-coverts, and often continues in this waste and unproductive state for half a century or more, till any portions of it are required for railroads, or for building upon, when a price equal to that of the best land in the district is often demanded and obtained. I know of thousands of acres in many parts of the south of England to which these remarks will apply, and if this is not obtaining land under false pretences—a legalized robbery of the poor for the aggrandizement of the rich, who were the law-makers—words have no meaning.

In this particular case the same course has been pursued. While I was writing these pages a friend was staying at Llandrindod, and I took the opportunity of asking him what was the present condition of the land more than sixty years after its inclosure. He informs me that, by inquiries among old inhabitants, he finds that at the time nothing was

done except to inclose the portions allotted to each landlord
with turf banks or other rough fencing; and that to this day
almost all the great boggy moor, with the mountain slopes
and summits, has not been improved in any way, either by
draining, cultivation, or planting, but is still wild, rough
pasture. But about thirty years after the inclosure the railway
from Shrewsbury through South Wales passed through the
place, and immediately afterwards a few villas and boarding-
houses were built, and some of the inclosed land was sold at
building prices. This has gone on year by year, and though
the resident population is still only about 2000, it is said that
10,000 visitors (more or less) come every summer, and the
chief increase of houses has been for their accommodation.
My friend tells me that, except close to the village and railway,
the whole country which was inclosed—many hundreds of
acres—is still bare and uncultivated, with hardly any animals
to be seen upon it. Milk is scanty and poor, and the only
butter is Cornish or Australian, so that the inclosure has not
led to the supply of the simplest agricultural needs of the
population. Even the piece of common that was reserved for
the use of the inhabitants is now used for golf-links!

Here, then, as in so many other cases, the express purpose
for which alone the legislature permitted the inclosure has
not been fulfilled, and in equity the whole of the land, and
the whole money proceeds of the sale of such portions as
have been built upon, should revert to the public. The prices

now realized by this almost worthless land, agriculturally, are enormous. In or near the village it sells for £1500 an acre, or even more, while quite outside these limits it is from £300 to £400.

In regard to this fundamental question of land ownership people are so blinded by custom and by the fact that it is sanctioned by the law, that it may be well for a moment to set these entirely on one side, and consider what would have been the proper, the equitable, and the most beneficial mode of dealing with our common and waste lands at the time of the last general Inclosure Act in the early years of the reign of Queen Victoria. Considering, then, that these uninclosed wastes were the last remnant of our country's land over which we, the public, had any opportunity of free passage to breathe pure air and enjoy the beauties of nature; considering that these wastes, although almost worthless agriculturally, were of especial value to the poor of the parishes or manors in which they were situated, not only giving them pasture for their few domestic animals, but in some cases peat for fuel and loppings of trees for fences or garden sticks; considering that an acre or two of such land, when inclosed and cultivated, would give them, in return for the labour of themselves and their families during spare hours, a considerable portion of their subsistence, would enable them to create a home from which they could not be ejected by the will of any landlord or employer, and would thus raise them at once to a condition of comparative independence and security, abolishing the terrible spectre of the workhouse for their old age, which now haunts the peasant or labourer throughout life, and is the fundamental cause of that exodus to the towns about which so much nonsense is talked; considering, further, that just in proportion as men rise in the social scale, these various uses of the waste lands become less and less vitally important, till, when we arrive at the country squire and great landowner, the only use of the inclosed common or moor is either to be used as a breeding-ground for game, or to add to some of his farms a few acres of land at an almost nominal rent;—considering all these circumstances, and further, that those who perform what is fundamentally the most important and the most beneficial of all work, the production of food, should be able to obtain at least the necessaries of life by that work, and secure a comfortable old age by their own fireside,—how would any lover of his country think that such lands *ought* to be dealt with in the best interests of the whole community?

Surely, that the very first thing to be done should be to provide that all workers upon the land, either directly or indirectly, should have plots of from one to five acres, in proportion to the amount of such waste and the needs of the inhabitants. The land thus allotted to be held by them in perpetuity, from the local authority, at a low rent such as any farmer would give for it as an addition to his farm. In cases where the amount of common land was very great in proportion to the population, some of the most suitable land might be reserved for a common pasture, for wood or fuel, or for recreation, and the remainder allotted to applicants from adjacent parishes where there was no common land.

Another thing that should be attended to in all such inclosures of waste land is the preservation for the people at large of rights of way over it in various directions, both to afford ample means of enjoying the beauties of nature and also to give pedestrians short cuts to villages, hamlets, or railway stations. One of the greatest blessings that might be easily attained if the land were resumed by the people to be held for the common good, would be the establishment of ample footpaths along every railway in the kingdom, with sufficient bridges or subways for safe crossing; and also (and more especially) along the banks of every river or brook, such paths to be diverted around any dwelling-house that may have gardens extending to the water's edge, all such paths to be made and kept in repair by the District Councils. Under the present system old paths are often closed, but we never hear of new ones being made, yet such are now more than ever necessary when most of our roads are rendered dangerous by motor-cars and cycles, and exceedingly disagreeable and unhealthy to pedestrians by the clouds of gritty dust continually raised by these vehicles.

This all-embracing system of land-robbery, for which

nothing is too great and nothing too small; which has absorbed meadow and forest, moor and mountain, which has appropriated most of our rivers and lakes and the fish that live in them; which often claims the very seashore and rocky coasts of our island home, fencing them off from the wayfarer who seeks the solace of their health-giving air and wild beauty, while making the peasant pay for his seaweed manure and the fisherman for his bait of shell-fish; which has desolated whole counties to replace men by sheep or cattle, and has destroyed fields and cottages to make a wilderness for deer and grouse; which has stolen the commons and filched the roadside wastes; which has driven the labouring poor into the cities, and has thus been the primary and chief cause of the lifelong misery, disease, and early death of thousands who might have lived lives of honest toil and comparative well-being had they been permitted free access to land in their native villages;—it is the advocates and beneficiaries of this inhuman system who, when a partial restitution of their unholy gains is proposed, are the loudest in their cries of "robbery"!

But all the robbery, all the spoliation, all the legal and illegal filching, has been on *their* side, and they still hold the stolen property. *They* made laws to legalize their actions, and, some day, we, the people, will make laws which will not only legalize but justify our process of restitution. It will justify it, because, unlike their laws, which always took from the poor to give to the rich—to the very class which made the laws—ours will only take from the superfluity of the rich, *not* to give to the poor or to any individuals, but to be so administered as to enable every man to live by honest work, to restore to the whole people their birthright in their native soil, and to relieve all alike from a heavy burden of unnecessary and unjust taxation. *This* will be the true statesmanship of the future, and it will be justified alike by equity, by ethics, and by religion.

In the preceding pages I have expressed the opinions which have been gradually formed as the result of the experience and study of my whole life. My first work on the subject was entitled "Land Nationalization: its Necessity and its Aims," and was published in the year 1882; and this, together with the various essays in the second volume of my "Studies Scientific and Social," published in 1900, may be taken as expressing the views I now hold, and as pointing out some of the fundamental conditions which I believe to be essential for the well-being of society.

But at the time of which I am now writing such ideas never entered my head. I certainly thought it a pity to inclose a wild, picturesque, boggy, and barren moor, but I took it for granted that there was *some* right and reason in it, instead of being, as it certainly was, both unjust, unwise, and cruel. But the surveying was interesting work, as every trickling stream, every tree, every mass of rock or boggy waterhole, had to be marked on the map in its true relative position, as well as the various footpaths or rough cart-roads that crossed the common in various directions.

At that time the medicinal springs, though they had been used from the time of the Romans, were only visited by a few Welsh or West of England people, and there was little accommodation for visitors, except in the small hotel where we lodged. One of our great luxuries here was the Welsh mutton fed on the neighbouring mountains, so small that a hind-quarter weighed only seven or eight pounds, but which, when hung a few days or a week, was most delicious eating. I agree with George Borrow in his praise of this dish. In his "Wild Wales" he says, "As for the leg of mutton it was truly wonderful; nothing so good had I ever tasted in the shape of a leg of mutton. The leg of mutton of Wales beats the leg of mutton of any other country, and I had never tasted a Welsh leg of mutton before. Certainly I shall never forget that first Welsh leg of mutton which I tasted, rich but delicate, replete with juices derived from the aromatic herbs of the noble Berwyn mountain, cooked to a turn, and weighing just four pounds." Well done, George Borrow! You had a good taste in ale and mutton, and were not afraid to acknowledge it.

CHAPTER VII

RESIDENCE IN SOUTH WALES: BRECKNOCKSHIRE AND GLAMORGANSHIRE

IT was in the summer or early autumn of 1841 that we left Kington for the tiny village of Trallong, a few miles beyond the town of Brecon, the parish we had to survey, and obtained lodgings in the house of a shoemaker, where we were very comfortable for some months. The house was pleasantly situated about two hundred and fifty feet above the river, with an uninterrupted view to the south-east over woody hills of moderate height to the fine range of the Great Forest, culminating in the double peaks of the Beacons, which were seen here fully separated with the narrow ridge connecting them. At sunset they were often beautifully tinted, and my brother made a charming little water-colour sketch of them, which, with most of his best sketches, were placed in an album by my sister, and this was stolen or lost while she was moving in London.

I looked daily at the Beacons with longing eyes, and on a fine autumn day one of the shoemaker's sons with a friend or two and myself started off to make the ascent. Though less than six miles from us in a straight line, we had to take a rather circuitous course over a range of hills, and then up to the head of a broad valley, which took us within a mile of the summit, making the distance about ten miles. But the day was gloriously fine, the country beautiful, and the view from the top very grand; while the summit itself was so curious as greatly to surprise me, though I did not fully appreciate its very instructive teaching till some years later, after I had ascended many other mountains, had studied Lyell's "Principles of Geology," and had fully grasped the modern views on sub-aërial denudation. As Brecknockshire is comparatively little known, and few English tourists make the ascent of the Beacons, a short account of them will be both interesting and instructive.

The northern face of the mountain is very rocky and precipitous, while on the southern and western sides easy slopes reach almost to the summit. The last few yards is, however, rather steep, and at the very top there is a thick layer of peat, which overhangs the rock a little. On surmounting this on the west side the visitor finds himself in a nearly flat triangular space, perhaps three or four acres in extent, bounded on the north by a very steep rocky slope, and on the other sides by steep but not difficult grass slopes. To the north-east he sees the chief summit about a quarter of a mile distant and nearly fifty feet higher, while connecting the two is a narrow ridge or saddle-back, which descends about a

THE BEACONS, LOOKING SOUTH.

hundred feet in a regular curve, and then rises again, giving an easy access to the higher peak. The top of this ridge is only a foot or two wide and very steep on the northern slope, but the southern slope is less precipitous, and about a hundred yards down it there is a small spring where the visitor can get deliciously cool and pure water. The north-eastern summit is also triangular, a little larger than the other, and bounded by a very dangerous precipice on the side towards Brecon, where there is a nearly vertical slope of craggy rock for three or four hundred feet and a very steep rocky slope for a thousand, so that a fall is almost certainly fatal, and several such accidents have occurred, especially when parties of young men from Brecon make a holiday picnic to the summit.

What strikes the observant eye as especially interesting is the circumstance that these two triangular patches, forming the culminating points of South Wales, both slope to the south-west, and by stooping down on either of them, and looking towards the other, we find that their surfaces correspond so closely in direction and amount of slope, that they impress one at once as being really portions of one continuous mountain summit. This becomes more certain when we look at the whole mountain mass, of which they form a part, known as the "Fforest Fawr," or great forest of Brecknock. This extends about twenty miles from east to west and ten or twelve miles from north to south; and in every part of it the chief summits are from 2000 to 2500 feet high, while near its western end, about twelve miles from the Beacons, is the second highest summit, Van Voel, reaching 2632 feet. Most of these mountains have rounded summits which are smooth and covered with grassy or sedgy vegetation, but many of them have some craggy slopes or precipices on their northern faces.

Almost the whole of this region is of the Old Red Sandstone formation, which here consists of nearly horizontal strata with a moderate dip to the south; and the whole of the very numerous valleys with generally smooth and gradually sloping sides which everywhere intersect it, must be all due to sub-aërial denudation—that is, to rain, frost, and snow—the *débris* due to which is carried away by the brooks and rivers. The geologist looks upon the rounded summits of these mountains as indications of an extensive gently undulating plateau, which had been slowly raised above the surface of the lakes or inland seas in which they had been deposited, and subjected to so little disturbance that the strata remain in a nearly horizontal position. When from the summit of any of these higher mountains we look over the wide parallel or radiating valleys with the rounded grassy ridges, and consider that the whole of the material that once filled all these valleys to the level of the mountain-top has been washed away day by day and year by year, by the very same agencies that after heavy rain now render turbid every brooklet, stream, and river, usually so clear and limpid, we obtain an excellent illustration of how Nature works in moulding the earth's surface by a process so slow as to be to us almost imperceptible.

This process of denudation is rendered especially clear to us by the singular formation of the twin summits of the Brecon Beacons. Here we are able, as it were, to catch

Nature at work. Owing to the rare occurrence of a nearly equal rate of denudation in four or five directions around this highest part of the original plateau, we have remaining for our inspection two little triangular patches of the original peat-covered surface joined together by the narrow saddle, as shown in the sketches opposite, giving a plan of the summits and a section through them to explain how accurately the two coincide in their slope with that of the original plateau.

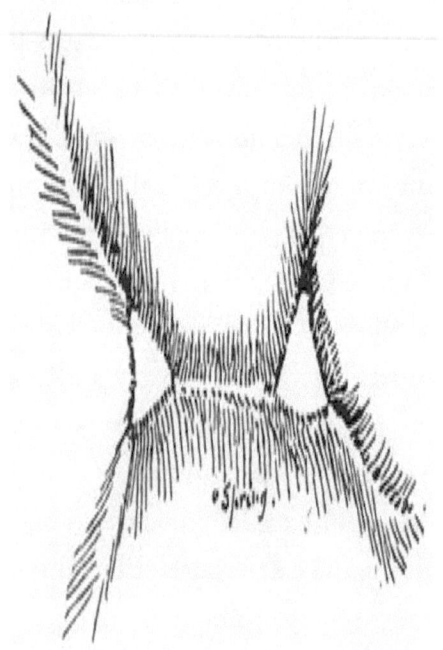

PLAN OF TOP OF BEACONS.

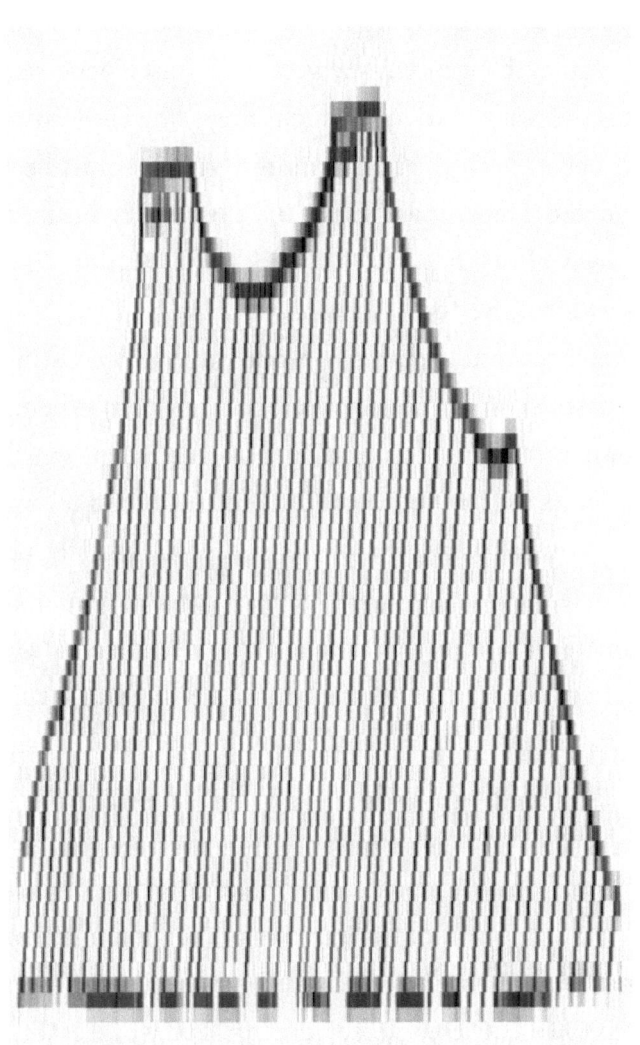

SECTION OF TOP OF BEACONS.

Every year the frost loosens the rock on the northern precipices, every heavy rain washes down earth from the ridge, while the gentler showers and mists penetrate the soil to the rock surface, which they slowly decompose. Thus, year by year, the flat portion of the summits becomes smaller, and a few thousand years will probably suffice to eat them away altogether, and leave rocky peaks more like that of Snowdon. The formation, as we now find it, is, in my experience, unique—that is, a mountain-top presenting two small patches of almost level ground, evidently being the last remnant of the great rolling plateau, out of which the whole range has been excavated. Double-headed mountains are by no means uncommon, but they are usually peaked or irregular, and carved out of inclined or twisted strata. The peculiarity of the Beacons consists in the strata being nearly horizontal and undisturbed, while the rock formation is not such as usually to break away into vertical precipices. The original surface must have had a very easy slope, while there were no meteorological conditions leading to great inequalities of weathering. The thick covering of peat has also aided in the result by preserving the original surface from being scored into gullies, and thus more rapidly denuded.

After we had completed most of our work at Trallong we had to go further up the valley to Devynock. This is an enormous parish of more than twenty thousand acres, divided into four townships or chapelries, the two eastern of which, Maescar and Senni, we had to survey. In these mountain districts, however, we only surveyed those small portions where the new roads or new inclosures had been made, the older maps being accepted as sufficiently accurate for the large uninclosed areas of mountain land. We first went to Senni Bridge, where both districts terminate in the Usk valley; but after a short time I went to stay in a little public-house at Senni in the midst of my work, while my brother stayed at Devy-nock or at Trallong, which latter was quite as near for half the work.

When I went up to Senni Street (Heol Senni, as it is called in Welsh) I greatly enjoyed wandering over the pretty valley which extended a long way into the mountains, flowing over nearly level meadows and with an unusually twisted course. This I found was so erroneously mapped, the numerous bends having been inserted at random as if of no importance, that I had to survey its course afresh. Above the village there were several lateral tributaries descending in deep woody dingles, often very picturesque, and these had usually one or more waterfalls in their course, or deep rocky chasms; and as these came upon me unexpectedly, and I had seen very few like them in Radnorshire, they were more especially attractive to me.

One Sunday afternoon I walked up the valley and over a mountain-ridge to the head waters of the Llia river, one of the tributaries of the river Neath, to see an ancient stone, named

Maen Llia on the ordnance map. I was much pleased to find a huge erect slab of Old Red Sandstone nearly twelve feet high, a photograph of which I am able to give through the kindness of Miss Florence Neale of Penarth. These strange relics of antiquity have always greatly interested me, and this, being the first I had ever seen, produced an impression which is still clear and vivid.

The people here were all thoroughly Welsh, but the landlord of the inn, and a young man who lived with him, spoke English fairly well.

MAEN LLIA.

Among the numerous Englishmen who visit Wales for business or pleasure, few are aware to what an extent this ancient British form of speech is still in use among the people, how many are still unable to speak English, and what an amount of poetry and legend their language contains. Some account of this literature is to be found in that very interesting book, George Borrow's "Wild Wales," and he claims for Dafydd ap Gwilym, a contemporary of our Chaucer, the position of "the greatest poetical genius that has appeared in Europe since the revival of literature." At the present day there are no less than twenty weekly newspapers and about the same number of monthly magazines published in the Welsh language, besides one quarterly and two bi-monthly reviews. Abstracts of the principal Acts of Parliament and Parliamentary papers are translated into Welsh, and one firm of booksellers, Messrs. Hughes and Son, of Wrexham, issue a list of more than three hundred Welsh books mostly published by themselves. Another indication of the wide use of the Welsh language and of the general education of the people, is the fact that the British and Foreign Bible Society now sell annually about 18,000 Bibles, 22,000 Testaments, and 10,000 special portions (as the Psalms, the Gospels, etc.); while the total sale of the Welsh Scriptures during the last century has been 3½ millions. Considering that the total population of Wales is only about 1½ millions, that two counties, Pembrokeshire and Radnorshire, do not speak

Welsh, and that the great seaports and the mining districts contain large numbers of English and foreign workmen, we have ample proof that the Welsh are still a distinct nation, with a peculiar language, literature, and history, and that the claim which they are now making for home rule, along with the other great subdivisions of the British Islands, is thoroughly justified.

It was late in the autumn of 1841 that we returned to Kington, and shortly after bade adieu to the wild but not very picturesque Radnorshire mountains for the more varied and interesting county of Glamorgan. I have no distinct recollection of our journey, but I believe it was by coach through Hay and Brecon to Merthyr Tydvil, and thence by chaise to Neath. One solitary example of the rhyming letters I used to write has been preserved, giving my younger brother Herbert an account of our journey, of the country, and of our work, of which, though very poor doggerel, a sample may be given. After a few references to family matters, I proceed to description.

"From Kington to this place we came
By many a spot of ancient fame,
But now of small renown,
O'er many a mountain dark and drear,
And vales whose groves the parting year
Had tinged with mellow brown;

And as the morning sun arose
New beauties round us to disclose,
We reached fair Brecon town;
Then crossed the Usk, my native stream,
A river clear and bright,
Which showed a fair and much-lov'd scene
Unto my lingering sight."

We had to go to Glamorganshire to partially survey and make a corrected map of the parish of Cadoxton-juxta-Neath, which occupies the whole northern side of the Neath valley from opposite the town of Neath to the boundary of the county at Pont-Nedd-Fychan, a distance of nearly fifteen miles, with a width varying from two to three miles, the boundary running for the most part along the crest of the mountains that bound the valley on the north-west. We lodged and boarded at a farmhouse called Bryncoch (Red Hill), situated on a rising ground about two miles north of the town. Here we stayed more than a year, living plainly but very well, and enjoying the luxuries of home-made bread, fresh butter and eggs, unlimited milk and cream, with cheese made from a mixture of cow's and sheep's milk, having a special flavour which I soon got very fond of. In this part of Wales it is the custom to milk the ewes chiefly for the purpose of making this cheese, which is very much esteemed.

A little rocky stream bordered by trees and bushes ran through the farm, and was one of my favourite haunts. There was one little sequestered pool about twenty feet long into which the water fell over a ledge about a foot high. This pool was seven or eight feet deep, but shallowed at the further end, and thus formed a delightful bathing-place. Ever since my early escape from drowning at Hertford, I had been rather shy of the water, and had not learned to swim; but here the distance was so short that I determined to try, and soon got to enjoy it so much that every fine warm day I used to go and plunge head first off my ledge and swim in five or six strokes to the shallow water. In this very limited sphere of action I gained some amount of confidence in the water, and afterwards should probably have been able to swim a dozen or twenty yards, so as to reach the bank of a moderate-sized river, or sustain myself till some neighbouring boat came to my assistance. But I have never needed even this moderate amount of effort to save my life, and have never had either the opportunity or inclination to become a practised swimmer. This was partly due to a physical deficiency which I was unable to overcome. My legs are unusually long for my height, and the bones are unusually large. The result is that they persistently sink in the water, bringing me into a nearly vertical position, and their weight renders it almost impossible to keep my mouth above water. This is the case even in salt water, and being also rather deficient in strength of muscle, I became disinclined to practise what I felt to be beyond my powers.

The parish being so extensive we had to stay at many different points for convenience of the survey, and one of these was about five miles up the Dulais valley, where we stayed at a small beershop in the hamlet of Crynant. I was often here alone for weeks together, and saw a good deal of the labourers and farmers, few of whom could speak any English. The landlady here brewed her own beer in very primitive fashion in a large iron pot or cauldron in the wash-house, and had it ready for sale in a few days—a rather thick and sweetish liquor, but very palatable. The malt and hops were bought in small quantities as wanted, and brewing took place weekly, or even oftener, when there was a brisk demand.

After living about a year at Bryn-coch we moved a little nearer the town to the other side of the Clydach river, and lodged with an old colliery surveyor, Samuel Osgood, in the employment of Mr. Price, of the Neath Abbey Iron Works. The house was an old but roomy cottage, and we had a large bedroom and a room downstairs for an office and living room, while Mr. Osgood had another, and there was also a roomy kitchen. A tramway from some collieries to the works ran in front of the house at a little distance, and we had a good view of the town and up the vale of Neath. Behind us rose the Drymau Mountain, nearly seven hundred feet above us, the level top of which was frequented by peewits, and whose steep slopes were covered with trees and bushes. Here we lived till I left Neath a year later, and were on the whole very comfortable, though our first experience was a rather trying one. The bedroom we occupied had been unused for years, and though it had been cleaned for our use we found that every part of it, bedstead, floor, and walls, in every crack and cranny, harboured the *Cimex lectularius,* or bedbug, which attacked us by hundreds, and altogether banished sleep. This required prompt and thorough measures, and my brother at once took them. I was sent to the town for some ounces of corrosive sublimate; the old wooden bedstead was taken to pieces, and, with the chairs, tables, drawers, etc., carried outside. The. poison was dissolved in a large pailful of water, and with this solution by means of a whitewasher's brush the whole of the floor was thoroughly soaked, so that the poison might penetrate every crevice, while the walls and ceiling were also washed over. The bedstead and furniture were all treated in the same way, and everything put back in its place by the evening. We did all the work ourselves, with the assistance of Mrs. Osgood and a servant-girl, and so effectual was the treatment that for nearly a year that we lived there we were wholly unmolested by insect enemies.

About that time the method of measuring the acreage of fields on maps by means of tracing-paper divided into squares of one chain each, with a beam-compass to sum up each line of squares, had recently come into use by surveyors; and Mr. Osgood amused himself by making a number of these compasses of various kinds of wood nicely finished and well polished, rather as examples of his skill than for any use he had for them, though he occasionally sold them to some of the local surveyors. He had these all suspended vertically on the wall instead of horizontally, as they are usually placed, and as they look best. While we were one day admiring the workmanship of an addition to the series, he remarked, "I dare say you don't know why I hang them up that way; very few people do." Of course, we acknowledged we did not know. "Well," said he, "it is very important. The air presses with a weight of fifteen pounds on every square inch, and if I hung them up level the pressure in the middle would very soon bend them, and they would be spoilt." My brother knew it was no good to try to show him his error, so he merely said, "Yes, that's a very good idea of yours," and left the old man in the happy belief that he was quite scientific in his methods.

After we had completed the survey and maps of Cadoxton, which occupied us about six months, we had not much to do except small pieces of work of various kinds. One of these was to make a survey and take soundings of the river between the bridge and the sea, a distance of three or four miles, for a proposed scheme of improving the navigation, making docks, etc., which was partly carried out some years later. We also had a little architectural and engineering work, in designing and superintending the erection of warehouses with powerful cranes, which gave me some insight into practical building. To assist in making working drawings and specifications, my brother had purchased a well-known work, Bartholomew's "Specifications for Practical Architecture." This book, though mainly on a very dry and technical subject, contained an introduction on the principles of Gothic architecture which gave me ideas upon the subject of the greatest interest and value, and which have enabled me to form an independent judgment on modern imitations of Gothic or of any other styles. Bartholomew was an enthusiast for Gothic, which he maintained was the only true and scientific system of architectural construction in existence. He showed how all the most striking and ornamental features of Gothic architecture are essential to the stability of a large stone-built structure—the lofty nave with its clerestory windows and arched roof; the lateral aisles at a lower level, also with arched roofs; the outer thrust of these arches supported by deep buttresses on the ground, with arched or flying buttresses above; and these again rendered more secure by being weighted down with rows of pinnacles, which add so much to the beauty of Gothic buildings. He rendered his argument more clear by giving a generalized cross-section of a cathedral, and drawing within the buttresses the figure of a man with outstretched arms pushing against the upper arches to resist their outward thrust, and being kept more steady by a heavy load upon his head and shoulders representing the pinnacle. This section and figure illuminated the whole construction of the masterpieces of the old architects so clearly and forcibly, that though I have not seen the book since, I have never forgotten it. It has furnished me with a standard by which to judge all architecture, and has guided my taste in such a small matter as the use of stone slabs over window openings in brick buildings, thus concealing the structural brick arch, and using stone as a beam, a purpose for which iron or wood is better suited. It also made me a very severe critic of modern imitations of Gothic in which we often see buttresses and pinnacles for ornament alone, when the roof is wholly of wood and there is no outward thrust to be guarded against; while in some cases we see useless gargoyles, which in the old buildings stretched out to carry the water clear of the walls, but which are still sometimes imitated when the water is carried into drains by iron gutters and water-pipes. I also learnt to appreciate the beautiful tracery of the large circular or pointed windows, whose harmonies and well-balanced curves and infinitely varied designs are a delight to the eye; while in most modern structures the attempts at imitating them are deplorable failures, being usually clumsy, unbalanced, and monotonous. One of the very few modern Gothic buildings in which the architect has caught the spirit of the old work is Barry's Houses of Parliament, which, whether in general effect or in its beautifully designed details, is a delight to the true lover of Gothic architecture.

CHAPTER VIII

SELF-EDUCATION IN SCIENCE AND LITERATURE

DURING the larger portion of my residence at Neath we had very little to do, and my brother was often away, either seeking employment or engaged upon small matters of business in various parts of the country. I was thus left a good deal to my own devices, and, having no friends of my own age, I occupied myself with various pursuits in which I had begun to take an interest. Having learnt the use of the sextant in surveying, and my brother having a book on Nautical Astronomy, I practised a few of the simpler observations. Among these were determining the meridian by equal altitudes of the sun, and also by the pole-star at its upper or lower culmination; finding the latitude by the meridian altitude of the sun, or of some of the principal stars; and making a rude sundial by erecting a gnomon towards the pole. For these simple calculations I had Hannay and Dietrichsen's Almanac, a copious publication which gave all the important data in the Nautical Almanac, besides much other interesting matter, useful for the astronomical amateur or the ordinary navigator. I also tried to make a telescope by purchasing a lens of about two feet focus at an optician's in Swansea, fixing it in a paper tube and using the eye-piece of a small opera-glass. With it I was able to observe the moon and Jupiter's satellites, and some of the larger star-clusters; but, of course, very imperfectly. Yet it served to increase my interest in astronomy, and to induce me to study with some care the various methods of construction of the more important astronomical instruments; and it also led me throughout my life to be deeply interested in the grand onward march of astronomical discovery.

But what occupied me chiefly and became more and more the solace and delight of my lonely rambles among the moors and mountains, was my first introduction to the variety, the beauty, and the mystery of Nature as manifested in the vegetable kingdom.

I obtained a shilling paper-covered book published by the Society for the Diffusion of Useful Knowledge, the title of which I forget, but which contained an outline of the structure of plants and a short description of their various parts and organs; and also a good description of about a dozen of the most common of the natural orders of British plants. This little book was a revelation to me, and for a year was my constant companion. On Sundays I would stroll in the fields and woods, learning the various parts and organs of any flowers I could gather, and then trying how many of them belonged to any of the orders described in my book. Great was my delight when I found that I could identify a Crucifer, an Umbellifer, and a Labiate; and as one after another the different orders were recognized, I began to realize for the first time the system that underlay all the variety of nature. When my brother was away and there was no work to do, I would spend the greater part of the day wandering over the hills or by the streams gathering flowers, and either determining their position from my book, or coming to the conclusion that they belonged to other orders of which I knew nothing, and as time went on I found that there were a very large number of these, including many of our most beautiful and curious flowers, and I felt that I *must* get some other book by which I could learn something about these also.

At length I obtained Lindley's "Elements of Botany," which to my disappointment did not contain references to British plants, and did not state which orders contained British species. The woodcuts and descriptions of all the natural orders were, however, very useful, and on the broad margins of the pages, I copied the characters of the British species from Loudon's "Encyclopædia of Plants," which I borrowed from a friend. I was thus able to identify most of the plants I met with.

But I soon found that by merely identifying the plants I gathered in my walks I lost much time in examining the same species several times, and even then not being always quite sure that I had found the same plant before. I therefore began to form a herbarium, collecting good specimens and drying them carefully between drying papers and a couple of boards weighted with books or stones. My brother, however, did not approve of my devotion to this study, even though I had absolutely nothing else to do, nor did he suggest any way in which I could employ my leisure more profitably. He said very little to me on the subject beyond a casual remark, but a letter from my mother showed me that he thought I was wasting my time. Neither he nor I could foresee that it would have any effect on my future life, and I myself only looked upon it as an intensely interesting occupation for time that would otherwise be wasted. Even when we were busy I had Sundays perfectly free, and used then to take long walks over the mountains with my collecting-box, which I brought home full of treasures. I first named the species as nearly as I could do so, and then laid them out to be pressed and dried. At such times I experienced the joy which every discovery of a new form of life gives to the lover of nature, almost equal to those raptures which I afterwards felt at every capture of new butterflies on the Amazon, or at the constant stream of new species of birds, beetles, and butterflies in Borneo, the Moluccas, and the Aru Islands.

It must be remembered that my ignorance of plants at this time was extreme. I knew the wild rose, bramble, hawthorn, buttercup, poppy, daisy, and foxglove, and a very few others equally common and popular, and this was all. I knew nothing whatever as to genera and species, nor of the large numbers of distinct forms related to each other and grouped into natural orders. My delight, therefore, was great when I was now able to identify the charming little eye-bright, the strange-looking cow-wheat and louse-wort, the handsome mullein and the pretty creeping toadflax, and to find that all of them, as well as the lordly foxglove, formed parts of one great natural order, and that under all their superficial diversity of form there was a similarity of structure which, when once clearly understood, enabled me to locate each fresh species with greater ease. The Crucifers, the Pea tribe, the Umbelliferæ, the Compositæ, and the Labiates offered great difficulties, and it was only after repeated efforts that I was able to name with certainty a few of the species, after which each additional discovery became a little less difficult, though the time I gave to the study before I left England was not sufficient for me to acquaint myself with more than a moderate proportion of the names of the species I collected.

Now, I have some reason to believe that this was the turning-point of my life, the tide that carried me on, not to fortune but to whatever reputation I have acquired, and which has certainly been to me a never-failing source of much health of body and supreme mental enjoyment. If my brother had had constant work for me so that I never had an idle day, and if I had continued to be similarly employed after I became of age, I should most probably have become entirely absorbed in my profession, which, in its various departments, I always found extremely interesting, and should therefore not have felt the need of any other occupation or study.

I know now, though I was ignorant of it at the time, that my brother's life was a very anxious one, that the difficulty of finding remunerative work was very great, and that he was often hard pressed to earn enough to keep us both in the very humble way in which we lived. He never alluded to this that I can remember, nor did I ever hear how much our board and lodging cost him, nor ever saw him make the weekly or monthly payments. During the seven years I was with him I hardly ever had more than a few shillings for personal expenses; but every year or two, when I went home, what new clothes were absolutely necessary were provided for me, with perhaps ten shillings or a pound as pocket-money till my next visit, and this, I think, was partly or wholly paid out of the small legacy left me by my grandfather. This seemed very hard at the time, but I now see clearly that even this was useful to me, and was really an important factor in moulding my character and determining my work in life. Had my father been a moderately rich man and had supplied me with a good wardrobe and ample pocket-money; had my brother obtained a partnership in some firm in a populous town or city, or had established himself in his profession, I might never have turned to nature as the solace and enjoyment of my solitary hours, my whole life would have been differently shaped, and though I should, no doubt, have given some attention to science, it seems very unlikely that I should have ever undertaken what at that time seemed rather a wild scheme, a journey to the almost unknown forests of the Amazon in order to observe nature and make a living by collecting. All this may have been pure chance, as I long thought it was, but of late years I am more inclined to Hamlet's belief when he said—

> "There's a divinity that shapes our ends,
> Rough-hew them how we will."

Of course, I do not adopt the view that each man's life, in all its details, is guided by the Deity for His special ends. That would be, indeed, to make us all conscious automata, puppets in the hands of an all-powerful destiny. But I have good reasons for the belief that we are surrounded by a host of unseen friends and relatives who have gone before us, and who have certain limited powers of influencing, and even, in particular cases, almost of determining, the actions of living persons, and may thus in a great variety of indirect ways modify the circumstances and character of any one or more individuals in whom they are specially interested. But a great number of these occurrences in every one's life are apparently what we term chance, and even if all are so, the conclusion I wish to lay stress upon is not affected. It is, that many of the conditions and circumstances that constitute our environment, though at the time they may seem unfortunate or even unjust, yet are often more truly beneficial than those which we should consider more favourable. Sometimes they only aid in the formation of character; sometimes they also lead to action which gives scope for the use of what might have been dormant or unused faculties (as, I think, has occurred in my own case); but much more frequently they seem to us wholly injurious, leading to a life of misery or crime, and turning what in themselves are good faculties to evil purposes. When this occurs in any large number of cases, as it certainly does with us now, we may be sure that it is the system of society that is at fault, and the most strenuous efforts of all who see this should be devoted, not to the mere temporary alleviation of the evils due to it, but to the gradual modification of the system itself. This is my present view. At the time of which I am now writing, I had not begun even to think of these matters, although facts which I now see to be of great importance in connection with them were being slowly accumulated for use in after-years.

It was during the time that I was most occupied out of doors with the observation and collection of plants that I began to write down, more or less systematically, my ideas on various subjects that interested me. Three of these early attempts have been preserved and are now before me. They all bear dates of the autumn or winter of 1843, when I was between nineteen and twenty years of age.

One of these is a rough sketch of a popular lecture on Botany, addressed to an audience supposed to be as ignorant as I was myself when I began to observe our native flowers. I was led to write it, partly on account of the difficulties I myself had felt in obtaining the kind of information I required, but chiefly on account of a lecture I had attended at Neath by a local botanist of some repute, which seemed to me so meagre, so uninteresting, and so utterly unlike what such a lecture ought to be, that I wanted to try if I could not do something better. The lecture in question consisted in an enumeration of the whole series of the "Linnæan Classes and Orders," stating their characters and naming a few of the plants comprised in each. It was illustrated by a series of coloured figures on cards about the size of ordinary playing cards, which the lecturer held up one after the other to show what he was talking about. The Linnæan system was upheld as being far the most useful as a means of determining the names of plants, and the natural system was treated as quite useless for beginners, and only suited for experienced botanists.

All this was so entirely opposed to views I had already formed, that I devoted a large portion of my lecture to the question of classification in general, showed that *any* classification, however artificial, was better than none, and that Linnaeus made a great advance when he substituted generic and specific names for the short Latin descriptions of species before used, and by classifying all known plants by means of a few well-marked and easily observed characters. I then showed how and why this classification was only occasionally, and as it were accidentally, a natural one; that in a vast number of cases it grouped together plants which were essentially unlike each other; and that for all purposes, except the naming of species, it was both useless and inconvenient. I then showed what the natural system of classification really was, what it aimed at, and the much greater interest it gave to the study of botany. I explained the principles on which the various natural orders were founded, and showed how often they gave us a clue to the properties of large groups of species, and enabled us to detect real affinities under very diverse external forms.

I concluded by passing in review some of the best marked orders as illustrating these various features. Although crudely written and containing some errors, I still think it would serve as a useful lecture to an audience generally ignorant of the whole subject, such as the young mechanics of a manufacturing town. Its chief interest to me now is that it shows my early bent towards classification, not the highly elaborate type that seeks to divide and subdivide under different headings with technical names, rendering the whole scheme difficult to comprehend, and being in most cases a hindrance rather than an aid to the learner, but a simple and intelligible classification which recognizes and defines all great natural groups, and does not needlessly multiply them on account of minute technical differences. It has always seemed to me that the natural orders of flowering plants afford one of the best, if not the very best, example of such a classification.

It is this attraction to classification, not as a metaphysically complete system, but as an aid to the comprehension of a subject, which is, I think, one of the chief causes of the success of my books, in almost all of which I have aimed at a simple and intelligible rather than a strictly logical arrangement of the subject-matter.

Another lecture, the draft for which I prepared pretty fully, was on a rather wider subject—"The Advantages of Varied Knowledge"—in opposition to the idea that it was better to learn one subject thoroughly than to know something of many subjects. In the case of a business or profession, something may be said for the latter view, but I treated it as a purely personal matter which led to the cultivation of a variety of faculties, and gave pleasurable occupation throughout life. A few extracts may, perhaps, be permitted from this early attempt. Speaking of a general acquaintance with history, biography, art, and science, I say, "There is an intrinsic value to ourselves in these varied branches of knowledge, so much indescribable pleasure in their possession, so much do they add to the enjoyment of every moment of our existence, that it is impossible to estimate their value, and we would hardly accept boundless wealth, at the cost, if it were possible, of their irrecoverable loss. And if it is thus we feel as to our general store of mental acquirements, still more do we appreciate the value of any particular branch of study we may ardently pursue. What pleasure would remain for the enthusiastic artist, were he forbidden to gaze upon the face of Nature, and transfer her loveliest scenes to his canvas? or for the poet, were the means denied him to rescue from oblivion the passing visions of his imagination? or to the chemist, were he snatched from his laboratory ere some novel experiment were concluded, or some ardently pursued theory confirmed? or to any of us, were we compelled to forego some intellectual pursuit that was bound up with our every thought? And here we see the advantage possessed by him whose studies have been in various directions, and who at different times has had many different pursuits, for whatever may happen, he will always find something in his surroundings to interest and instruct him."

And further on, as illustrations of the interest in common things conferred by a knowledge of the elementary laws of physical science, I remark—

"Many who marvel at the rolling thunder care not to inquire what causes the sound which is heard when a tightly-fitting cork is quickly drawn from a bottle, or when a whip is cracked or a pistol fired; and while they are struck with awe and admiration at the dazzling lightning, look upon the sparks drawn from a cat's back on a frosty evening and the slight crackle that accompanies them as being only fit to amuse a child; yet in each case the cause of the trifling and of the grand phenomena is the same. He who has extended his inquiries into the varied phenomena of nature learns to despise no fact, however small, and to consider the most apparently insignificant and common occurrences as much in need of explanation as those of a grander and more imposing character. He sees in every dewdrop trembling on the grass causes at work analogous to those which have produced the spherical figure of the earth and planets; and in the beautiful forms of crystallization on his window-panes on a frosty morning he recognizes the action of laws which may also have a part in the production of the similar forms of plants and of many of the lower animal types. Thus the simplest facts of everyday life have to him an inner meaning, and he sees that they depend upon the same general laws as those that are at work in the grandest phenomena of nature."

I then pass in review the chief arts and sciences, showing their inter-relations and unsolved problems; and in remarking on the Daguerrotype, then the only mode of photographic portraiture, I make a suggestion that, though very simple, has not yet been carried out. It is as follows:— "It would be a curious and interesting thing to have a series of portraits taken of a person each successive year. These would show the gradual changes from childhood to old age in a very striking manner; and if a number of such series from different individuals were obtained, and a brief outline given of their lives during each preceding year, we should have materials not merely for the curious to gaze at, but which might elucidate the problem of how far the mind reacts upon the countenance. We should see the effects of pain or pleasure, of idleness or activity, of dissipation or study, and thus watch the action of the various passions of the mind in modifying the form of the body, and particularly the expression of the features."

Now that photography is so widespread and so greatly improved, it is rather curious that nothing of this kind has been done. Some of our numerous scientific societies might offer to take such photographs of any of their members who would agree to be taken regularly, and would undertake to have one or two of their children similarly taken till they came of age, and also to prepare a very short record each year of the main events or occupations of their lives. If this were widely done in every part of the country, a most interesting and instructive collection of those series which were most complete would be obtained. I have given the concluding passage of the lecture as it appears in the rough draft, which was never rewritten.

"Can we believe that we are fulfilling the purpose of our existence while so many of the wonders and beauties of the creation remain unnoticed around us? While so much of the mystery which man has been able to penetrate, however imperfectly, is still all dark to us? While so many of the laws which govern the universe and which influence our lives are, by us, unknown and uncared for? And this not because we want the power, but the will, to acquaint ourselves with them. Can we think it right that, with the key to so much that we ought to know, and that we should be the better for knowing, in our possession, we seek not to open the door, but allow this great store of mental wealth to lie unused, producing no return to us, while our highest powers and capacities rust for want of use?

"It is true that man is still, as he always has been, subject to error; his judgments are often incorrect, his beliefs false, his opinions changeable from age to age. But experience of error is the best guide to truth, often dearly bought, and, therefore, the more to be relied upon. And what is it but the accumulated experience of past ages that serves us as a beacon light to warn us from error, to guide us in the way of truth? How little should we know had the knowledge acquired by each preceding age died with it! How blindly should we grope our way in the same obscurity as did our ancestors, pursue the same phantoms, make the same fatal blunders, encounter the same perils, in order to purchase the same truths which had been already acquired by the same process, and lost again and again in bygone ages! But the wonder-working press prevents this loss; truths once acquired are treasured up by it for posterity, and each succeeding generation adds something to the stock of acquired knowledge, so that our acquaintance with the works of nature is ever increasing, the range of our inquiries is extended each age, the power of mind over matter becomes, year by year, more complete. Yet our horizon ever widens, the limits to our advance seem more distant than ever, and there seems nothing too noble, too exalted, too marvellous, for the ever-increasing knowledge of future generations to attain to.

"Is it not fitting that, as intellectual beings with such high powers, we should each of us acquire a knowledge of what past generations have taught us, so that, should the opportunity occur, we may be able to add somewhat, however small, to the fund of instruction for posterity? Shall we not, then, feel the satisfaction of having done all in our power to improve by culture those higher faculties that distinguish us from the brutes, that none of the talents with which we may have been gifted have been suffered to lie altogether idle? And, lastly, can any reflecting mind have a doubt that, by improving to the utmost the nobler faculties of our nature in this world, we shall be the better fitted to enter upon and enjoy whatever new state of being the future may have in store for us?"

These platitudes are of no particular interest, except as snowing the bent of my mind at that period, and as indicating a disposition for discursive reading and study, which has been a great advantage to myself, and which has enabled me to write on a variety of subjects without committing any very grievous blunders (so far as my critics have pointed out), and with, I hope, some little profit to my readers.

In April, 1843, my father died at Hoddesdon, at the age of seventy-two, and was buried in the family vault in St. Andrew's Churchyard, Hertford. As my sister's school was not paying very well, and it was necessary to economize as much as possible, the house was given up early the following year, my mother took an engagement as housekeeper in a gentleman's family at Isleworth, and my sister obtained a post as teacher at an Episcopal College, then just founded by the Bishop of Georgia (Dr. Elliott), at Montpelier Springs, seventeen miles from Macon, and left England in August, 1844.

Shortly before I came of age in January, 1844, my brother told me that as he had no work in prospect it was necessary that I should leave him and look out for myself; so I determined to go up to London and endeavour to obtain some employment.

As the period of my home and school life and subsequent

tutelage under my brother now came to an end, and I had for the future to make my own way in the world, this affords a suitable occasion for a brief review of the chief points in my character, which may now be considered to have been fairly determined, although some portions of it had not yet had opportunity for full development.

I do not think that at this time I could be said to have shown special superiority in any of the higher mental faculties, but I possessed a strong desire to know the causes of things, a great love of beauty in form and colour, and a considerable but not excessive desire for order and arrangement in whatever I had to do. If I had one distinct mental faculty more prominent than another, it was the power of correct reasoning from a review of the known facts in any case to the causes or laws which produced them, and also in detecting fallacies in the reasoning of other persons. This power has greatly helped me in all my writings, especially those on natural history and sociology. The determination of the direction in which I should use these powers was due to my possession in a high degree of the two mental qualities usually termed emotional or moral, an intense appreciation of the beauty, harmony, and variety in nature and in all natural phenomena, and an equally strong passion for justice as between man and man— an abhorrence of all tyranny, all compulsion, all unnecessary interference with the liberty of others. These characteristics, combined with certain favourable conditions, some of which have already been referred to, have determined the direction of the pursuits and inquiries in which I have spent a large portion of my life.

It will be well to state here certain marked deficiencies in my mental equipment which have also had a share in determining the direction of my special activities. My greatest, though not perhaps most important, defect is my inability to perceive the niceties of melody and harmony in music; in common language, I have no ear for music. But as I have a fair appreciation of time, expression, and general harmony, I am deeply affected by grand, pathetic, or religious music, and can at once tell when the heart and soul of the musician is in his performance, though any number of technical errors, false notes, or harsh discords would pass unnoticed. Another and more serious defect is in verbal memory, which, combined with the inability to reproduce vocal sounds, has rendered the acquirement of all foreign languages very difficult and distasteful. This, with my very imperfect school training, added to my shyness and want of confidence, must have caused me to appear a very dull, ignorant, and uneducated person to numbers of chance acquaintances. This deficiency has also put me at a great disadvantage as a public speaker. I can rarely find the right word or expression to enforce or illustrate my argument, and constantly feel the same difficulty in private conversation. In writing it is not so injurious, for when I have time for deliberate thought

I can generally express myself with tolerable clearness and accuracy. I think, too, that the absence of the flow of words which so many writers possess has caused me to avoid that extreme diffuseness and verbosity which is so great a fault in many scientific and philosophical works.

Another important defect is in the power of rapidly seeing analogies or hidden resemblances and incongruities, a deficiency which, in combination with that of language, has produced the total absence of wit or humour, paradox or brilliancy, in my writings, although no one can enjoy and admire these qualities more than I do. The rhythm and pathos, as well as the inimitable puns of Hood, were the delight of my youth, as are the more recondite and fantastic humour of Mark Twain and Lewis Carroll in my old age. The faculty which gives to its possessor wit or humour is also essential to the high mathematician, who is almost always witty or poetical as well; and I was therefore debarred from any hope of success in this direction; while my very limited power of drawing or perception of the intricacies of form were equally antagonistic to much progress as an artist or a geometrician.

Other deficiencies of great influence in my life have been my want of assertiveness and of physical courage, which, combined with delicacy of the nervous system and of bodily constitution, and a general disinclination to much exertion, physical or mental, have caused that shyness, reticence, and love of solitude which, though often misunderstood and leading to unpleasant results, have, perhaps, on the whole, been beneficial to me. They have helped to give me those long periods, both at home and abroad, when, alone and surrounded only by wild nature and uncultured man, I could ponder at leisure on the various matters that interested me. Thus was induced a receptiveness of mind which enabled me at various times to utilize what appeared to me as sudden intuitions—flashes of light leading to a solution of some problem which was then before me; and these flashes would often come to me when, pen in hand, I was engaged in writing on a subject on which I had no intention or expectation of saying anything new.

Before leaving this sketch of my mental nature at the threshold of my uncontrolled life, I may properly say a few words on the position I had arrived at in regard to the great question of religious belief. I have already shown that my early home training was in a thoroughly religious but by no means rigid family, where, however, no religious doubts were ever expressed, and where the word "atheist" was used with bated breath as pertaining to a being too debased almost for human society. The only regular teaching I received was to say or hear a formal prayer before going to bed, hearing grace before and after dinner, and learning a collect every Sunday morning, the latter certainly one of the most stupid ways of inculcating religion ever conceived. On Sunday evenings, if we did not go to church or chapel, my father would read some old sermon, and when we did go we were asked on our return what was the text. The only books allowed to be read on Sundays were the "Pilgrim's Progress" or "Paradise Lost," or some religious tracts or moral tales, or the more interesting parts of the Bible were read by my mother, or we read ourselves about Esther and Mordecai or Bel and the Dragon, which were as good as any story-book. But all this made little impression upon me, as it never dealt sufficiently with the mystery, the greatness, the ideal and emotional aspects of religion, which only appealed to me occasionally in some of the grander psalms and hymns, or through the words of some preacher more impassioned than usual.

As might have been expected, therefore, what little religious belief I had very quickly vanished under the influence of philosophical or scientific scepticism. This came first upon me when I spent a month or two in London with my brother John, as already related; and during the seven years I lived with my brother William, though the subject of religion was not often mentioned, there was a pervading spirit of scepticism, or free-thought as it was then called, which strengthened and confirmed my doubts as to the truth or value of all ordinary religious teaching.

He occasionally borrowed interesting books which I usually read. One of these was an old edition of Rabelais' works, which both interested and greatly amused me; but that which bears most upon the present subject was a reprint of lectures on Strauss' "Life of Jesus," which had not then been translated into English. These lectures were, I think, delivered by some Unitarian minister or writer, and they gave an admirable and most interesting summary of the whole work. The now well-known argument, that all the miracles related in the Gospels were mere myths, which in periods of ignorance and credulity always grow up around all great men, and especially around all great moral teachers when the actual witnesses of his career are gone and his disciples begin to write about him, was set forth with great skill. This argument appeared conclusive to my brother and some of his friends with whom he discussed it, and, of course, in my then frame of mind it seemed equally conclusive to me, and helped to complete the destruction of whatever religious beliefs still lingered in my mind. It was not till many years afterwards that I saw reason to doubt this whole argument, and to perceive that it was based upon pure assumptions which were not in accordance with admitted historical facts.

My brother never went to church himself, but for the first few years I was with him he sent me once every Sunday; but, of course, the only effect of this was to deepen my spirit of scepticism, as I found no attempt in any of the clergymen to reason on any of the fundamental questions at the root of the Christian and every other religion. Many of our acquaintances were either church- or chapel-goers, but usually as a matter of form and convention, and, on the whole, religion seemed to have no influence whatever on their conduct or conversation. The majority, especially of the younger men, were either professors of religion who thought or cared nothing about it, or were open sceptics and scorners.

In addition to these influences my growing taste for various branches of physical science and my increasing love of nature disinclined me more and more for either the observances or the doctrines of orthodox religion, so that by the time I came of age I was absolutely non-religious, I cared and thought nothing about it, and could be best described by the modern term "agnostic."

The next four years of my life were also of great importance both in determining the direction of my activity, and in laying the foundation for my study of special subjects through which I have obtained most admiration or notoriety. These will form the subject of the following chapter.

CHAPTER IX

LIFE AT LEICESTER AND NEATH

As I came of age in January, 1844, and there was nothing doing at Neath, I left my brother about the middle of December so as to spend the Christmas with my mother and sister at Hoddesdon, after which I returned to London, sharing my brother John's lodging till I could find some employment. At that time the tithe-commutation surveys were nearly all completed, and the rush of railway work had not begun: surveying was consequently very slack. As my brother William, who had a large acquaintance among surveyors and engineers all over the south of England, could not find employment, except some very small local business, I felt it to be quite useless for me to seek for similar work. I therefore determined to try for some post in a school to teach English, surveying, elementary drawing, etc. Through some school agency I heard of two vacancies that might possibly suit. The first required, in addition to English, junior Latin and algebra. I applied for this, but failed to obtain the appointment, as neither my Latin nor my algebra was up to the required standard.

My next attempt was more hopeful, as drawing,

surveying, and mapping were required. On this occasion I met a clergyman, a young man, easy and friendly in his manner. I had taken with me a small coloured map I had made at Neath to serve as a specimen, and also one or two pencil sketches. These seemed to satisfy him, and as I was only wanted to take the junior classes in English reading, writing, and arithmetic, teach a very few boys surveying, and beginners in drawing, he agreed to engage me. I was to live in the house, preside over the evening preparation of the boarders (about twenty in number), and to have, I think, thirty or forty pounds a year, with which I was quite satisfied. My employer was the Rev. Abraham Hill, Headmaster of the Collegiate School at Leicester.

I commenced work in about a fortnight. After a few weeks, finding I knew a little Latin, Mr. Hill asked me to take the lowest class, and even that required some preparation in the evening. Mr. Hill was a good mathematician, and finding I was desirous of learning a little more algebra, offered to assist me. He lent me Hind's Algebra, which I worked all through successfully, and this was followed by the same author's Trigonometry, which I also went through, with occasional struggles. Then I attacked the Differential Calculus, and worked through that; but I could never fully grasp the essential principle of it. Finally, I began the Integral Calculus, and here I found myself at the end of my tether. I learnt some of the simpler processes, but very soon got baffled, and felt that I wanted some faculty necessary for seeing my way through what seemed to me an almost trackless labyrinth. Whether, under Mr. Hill's instruction, I should ultimately have been able to overcome these difficulties I cannot positively say, but I have good reason to believe that I never should have done so. Briefly stated, just as no amount of teaching or practice would ever have made me a good musician, so, however much time and study I gave to the subject, I could never have become a good mathematician. Whether all this work did me any good or not, I am rather doubtful. My after-life being directed to altogether different studies, I never had occasion to use my newly acquired knowledge, and soon forgot most of the processes. But it gave me an interest in mathematics which I have never lost; and I rarely come across a mathematical investigation without looking through it and trying to follow the reasoning, though I soon get lost in the formulæ. Still, the ever-growing complexity of the higher mathematics has a kind of fascination for me as exhibiting powers of the human mind so very far above my own.

There was in Leicester a very good town library, to which I had access on paying a small subscription, and as I had time for several hours' reading daily, I took full advantage of it. Among the works I read here, which influenced my future, were Humboldt's "Personal Narrative of Travels in South America," which was, I think, the first book that gave me a desire to visit the tropics. I also read here Prescott's "History of the Conquests of Mexico and Peru," Robertson's "History of Charles V." and his "History of America," and a number of other standard works. But perhaps the most important book I read was Malthus's "Principles of Population," which I greatly admired for its masterly summary of facts and logical induction to conclusions. It was the first work I had yet read treating of any of the problems of philosophical biology, and its main principles remained with me as a permanent possession, and twenty years later gave me the long-sought clue to the effective agent in the evolution of organic species.

It was at Leicester that I was first introduced to a subject which I had at that time never heard of, but which has played an important part in my mental growth—psychical research, as it is now termed. Some time in 1844 Mr. Spencer Hall gave some lectures on mesmerism illustrated by experiments, which I, as well as a few of the older boys, attended. I was greatly interested and astonished at the phenomena exhibited, in some cases with persons who volunteered from the audience; and I was also impressed by the manner of the lecturer, which was not at all that of the showman or the conjurer. At the conclusion of the course he assured us that most persons possessed in some degree the power of mesmerising others, and that by trying with a few of our younger friends or acquaintances, and simply doing what we had seen him do, we should probably succeed. He also showed us how to distinguish between the genuine mesmeric trance, and any attempt to imitate it.

In consequence of this statement, one or two of the elder boys tried to mesmerise some of the younger ones, and in a short time succeeded; and they asked me to see their experiments. I found that they could produce the trance state, which had all the appearance of being genuine, and also a cataleptic rigidity of the limbs by passes and by suggestion, both in the trance and afterwards in the normal waking state. This led me to try myself in the privacy of my own room, and I succeeded after one or two attempts in mesmerising three boys from twelve to sixteen years of age, while on others within the same ages I could produce no effect, or an exceedingly slight one. During the trance they seemed in a state of semi-torpor, with apparently no volition. They would remain perfectly quiescent so long as I did not notice them, but would at once answer any questions or do anything I told them. On the two boys with whom I continued to experiment for some time, I could produce catalepsy of any limb or of the whole body, and in this state they could do things which they could not, and certainly would not have done in their normal state. For example, on the rigid outstretched arm I would hang at the wrist an ordinary bedroom chair, and the boy would hold it there for several minutes, while I sat down and wrote a short letter for instance, without any complaint, or making any remark when I took it off. I never left it more than five minutes, because I was afraid that some injury might be caused by it. I soon found that this rigidity could be produced in those who had been mesmerised by suggestion only, and in this way often fixed them in any position, notwithstanding their efforts to change it. One experiment was to place a shilling on the table in front of a boy, and then say to him, "Now, you can't touch that shilling." He would at once move his hand towards it, but when half-way it would seem to stick fast, and all his efforts could not bring it nearer, though he was promised the shilling if he could take it.

But perhaps the most interesting group of phenomena to me were those termed phreno-mesmerism. I had read, when with my brother, George Combe's "Constitution of Man," with which I had been greatly interested, and afterwards one of the writer's works on Phrenology, and at the lecture I had seen some of the effects of exciting the phrenological organs by touching the corresponding parts of the patient's head. But as I had no book containing a chart of the organs, I bought a small phrenological bust to help me in determining the positions.

Having my patient in the trance, and standing close to him, with the bust on my table behind him, I touched successively several of the organs, the position of which it was easy to determine. After a few seconds he would change his attitude and the expression of his face in correspondence with the organ excited. In most cases the effect was unmistakable, and superior to that which the most finished actor could give to a character exhibiting the same passion or emotion.

At this very time the interest excited by painless surgical operations during the mesmeric trance was at its height, as I have described in my "Wonderful Century" (chap, xxi.), and I had read a good deal about these, and also about the supposed excitement of the phrenological organs, and the theory that these latter were caused by mental suggestion from the operator to the patient, or what is now termed telepathy. But as the manifestations often occurred in a different form from what I expected, I felt sure that this theory was not correct. One day I intended to touch a particular organ, and the effect on the patient was quite different from what I expected, and looking at the bust while my finger was still on the boy's head, I found that I was not touching the part I supposed, but an adjacent part, and that the effect exactly corresponded to the organ touched and not to the organ I *thought* I had touched, completely disproving the theory of suggestion. I then tried several experiments by looking away from the boy's head while I put my finger on it at random, when I always found that the effect produced corresponded to that indicated by the bust. I thus established, to my own satisfaction, the fact that a real effect was produced on the actions and speech of a mesmeric patient by the operator touching various parts of the head; that the effect corresponded with the natural expression of the emotion due to the phrenological organ situated at that part—as combativeness, acquisitiveness, fear, veneration, wonder, tune, and many others; and that it was in no way caused by the will or suggestion of the operator.

As soon as I found that these experiments were successful, I informed Mr. Hill, who made no objection to my continuing them, and several times came to see them. He was so much impressed that one evening he invited two or three friends who were interested in the subject, and with my best patient I showed most of the phenomena.

The importance of these experiments to me was that they convinced me, once for all, that the antecedently incredible may nevertheless be true; and, further, that the accusations of imposture by scientific men should have no weight whatever against the detailed observations and statements of other men, presumably as sane and sensible as their opponents, who had witnessed and tested the phenomena, as I had done myself in the case of some of them.

While living at Leicester I first met Henry Walter Bates. How I was introduced to him I do not exactly remember, but I rather think I heard him mentioned as an enthusiastic entomologist, and met him at the library. I found that his specialty was beetle collecting, though he also had a good set of British butterflies. Of the former I knew nothing, but as I already knew the fascinations of plant life I was quite prepared to take an interest in any other department of nature. He asked me to see his collection, and I was amazed to find the great number and variety of beetles, their many strange forms and often beautiful markings or colouring, and was even more surprised when I found that almost all I saw had been collected around Leicester, and that there were still many more to be discovered. If I had been asked before how many different kinds of beetles were to be found in any small district near a town, I should probably have guessed fifty or at the outside a hundred, and thought that a very liberal allowance. But I now learnt that many hundreds could easily be collected, and that there were probably a thousand different kinds within ten miles of the town; and he showed me a thick volume containing descriptions of more than three thousand species inhabiting the British Isles. I also learnt from him in what an infinite variety of places beetles may be found, while some may be collected all the year round, so I at once determined to begin collecting, as I did not find a great many new plants near Leicester. I therefore obtained a collecting bottle, pins, and a store-box; and in order to learn their names and classification I obtained, at wholesale price through Mr. Hill's bookseller, Stephen's "Manual of British Coleoptera," which henceforth for some years gave me almost as much pleasure as Lindley's "Botany," with my MSS. descriptions, had already done.

This new pursuit gave a fresh interest to my Wednesday and Saturday afternoon walks into the country, when two or three of the boys often accompanied me. The most delightful of all our walks was to Bradgate Park, about five miles from the town, a wild, neglected park with the ruins of a mansion, and many fine trees and woods and ferny or bushy slopes. Sometimes the whole school went for a picnic, the park at that time being quite open, and we hardly ever met any one. After we got out of the town there was a wide grassy lane that led to it, which itself was a delightful walk and was a good collecting ground for both plants and insects. For variety we had the meadows along the course of the little river Soar, which were very pleasant in spring and summer. Twice during the summer the whole of the boarders were taken for a long day's excursion. The first time we went to Kenilworth Castle, about thirty miles distant, driving in coaches by pleasant country roads, and passing through Coventry. Towards the autumn we had a much longer excursion, partly by coach and partly by canal boat, to a very picturesque country with wooded hills and limestone cliffs, rural villages, and an isolated hill, from the top of which we had a very fine and extensive view. I think it must have been in Derbyshire, near Wirksworth, as there is a long canal tunnel on the way there. One of the rough out-of-door sketches made on this occasion is reproduced here on a reduced scale, as well as a more finished drawing of some village, perhaps near Leicester, as they may possibly enable some reader to recognize the localities, and also serve to show the limits of my power as an artist.

IN DERBYSHIRE

IN LEICESTERSHIRE.

Early in the year 1846 I received the totally unexpected news of the death of my brother William at Neath. He had been in London to give evidence before a committee on the South Wales Railway Bill, and returning at night caught a severe cold by being chilled in a wretched third-class carriage, succeeded by a damp bed at Bristol. This brought on congestion of the lungs, to which he speedily succumbed. I and my brother John went down to Neath to the funeral, and as William had died without a will, we had to take out letters of administration. Finding from my brother's papers that he had obtained a small local business, and that there was railway work in prospect, I determined to take his place, and at once asked permission of Mr. Hill to be allowed to leave at Easter.

My two years spent at Leicester had been in many ways useful to me, and had also a determining influence on my whole future life. It satisfied me that I had no vocation for teaching, for though I performed my duties I believe quite to Mr. Hill's satisfaction, I felt myself out of place, partly because I knew no subject—with the one exception of surveying—sufficiently well to be able to teach it properly, but mainly because a completely subordinate position was distasteful to me, although I could not have had a more considerate employer than Mr. Hill. The time and opportunity I had for reading was a great advantage, and gave me an enduring love of good literature. The events which formed a turning-point in my life were, first, my acquaintance with Bates, and through him deriving a taste for the wonders of insect-life, opening to me a new aspect of nature, and later on finding in him a companion without whom I might never have ventured on my journey to the Amazon. The other and equally important circumstance was my reading Malthus, without which work I should probably not have hit upon the theory of natural selection and obtained full credit for its independent discovery. My two years spent at Leicester must, therefore, be considered as perhaps the most important in my early life.

At Easter I bade farewell to Leicester and went to Neath with my brother John, in order to wind up our brother William's affairs. We found from his books that a considerable amount was owing to him for work done during the past year or two, and we duly made out accounts of all these and sent them in to the respective parties. Some were paid at once, others we had to write again for and had some trouble to get paid.

When we had wound up William's affairs as well as we could, my brother John returned to London, and I was left to see if any work was to be had, and in the mean time devoted myself to collecting butterflies and beetles. While at Leicester I had been altogether out of the business world and do not remember even looking at a newspaper, or I might have heard something of the great railway mania which that year reached its culmination. I now first heard rumours of it, and some one told me of a civil engineer in Swansea who wanted all the surveyors he could get, and that they all had two guineas a day, and often more. This I could hardly credit, but I wrote to the gentleman, who soon after called on me, and asked me if I could do levelling. I told him I could, and had a very good level and levelling staves. After some little conversation he told me he wanted a line of levels up the Vale of Neath to Merthyr Tydfil for a proposed railway, with cross levels at frequent intervals, and that he would give me two guineas a day, and all expenses of chain and staff men, hotels, etc. He gave me all necessary instructions, and said he would send a surveyor to map the route at the same time. This was, I think, about midsummer, and I was hard at work till the autumn, and enjoyed myself immensely. It took me up the south-east side of the valley, of which I knew very little, along pleasant lanes and paths through woods and by streams, and up one of the wildest and most picturesque little glens I have ever explored. Here we had to climb over huge rocks as big as houses, ascend cascades, and take cross-levels up steep banks and precipices all densely wooded. It was surveying under difficulties, and excessively interesting. After the first rough levels were taken and the survey made, the engineers were able to mark out the line provisionally, and I then went over the actual line to enable the sections to be drawn as required by the Parliamentary

Standing Orders.

In this year of wild speculation it is said that plans and sections for 1263 new railways were duly deposited, having a proposed capital of £563,000,000, and the sum required to be deposited at the Board of Trade was so much larger than the total amount of gold in the Bank of England and notes in circulation at the time, that the public got frightened, a panic ensued, shares in the new lines which had been at a high premium fell almost to nothing, and even the established lines were greatly depreciated. Many of the lines were proposed merely for speculation, or to be bought off by opposing lines which had a better chance of success. The line we were at work on was a branch of the Great Western and South Wales Railway then making, and was for the purpose of bringing the coal and iron of Merthyr Tydfil and the surrounding district to Swansea, then the chief port of South Wales. But we had a competitor along the whole of our route in a great line from Swansea to Yarmouth, by way of Merthyr, Hereford, Worcester, and across the midland agricultural counties, called, I think, the East and West Junction Railway, which sounded grand, but which had no chance of passing.

It competed, however, with several other lines, and I heard that many of these agreed to make up a sum to buy off its opposition. Not one-tenth of the lines proposed that year were ever made, and the money wasted upon surveyors, engineers, and law expenses must have amounted to millions.

Finding it rather dull at Neath living by myself, I persuaded my brother to give up his work in London as a journeyman carpenter and join me, thinking that with his practical experience and my general knowledge, we might be able to do architectural, building, and engineering work, as well as surveying, and in time get up a profitable business. We returned together early in January, and continued to board and lodge with Mr. Sims in the main street, where I had been very comfortable, till the autumn, when, hearing that my sister would probably be home from America the following summer, and my mother wishing to live with us, we took a small cottage close to Llantwit Church, and less than a mile from the middle of the town. It had a nice little garden and yard, with fowl-house, shed, etc., going down to the Neath Canal, immediately beyond which was the river Neath, with a pretty view across the valley to Cadoxton and the fine Drumau mountain.

Having the canal close at hand and the river beyond, and then another canal to Swansea, made us long for a small boat, and not having much to do, my brother determined to build one, so light that it could be easily drawn or carried from the canal to the river, and so give access to Swansea. It was made as small and light as possible to carry two or, at most, three persons. When finished, we tried it with much anxiety, and found it rather unstable, but with a little ballast at the bottom and care in moving, it did very well, and was very easy to row. One day I persuaded my mother to let me row her to Swansea, where we made a few purchases; and then came back quite safely till within about a mile of home, when, passing under a bridge, my mother put her hand out to keep the boat from touching, and leaning over a little too much, the side went under water, and upset us both. As the water was only about two or three feet deep we escaped with a thorough wetting. The boat was soon baled dry, and then I rowed on to Neath Bridge, where my mother got out and walked home, and did not trust herself in our boat again, though I and my brother had many pleasant excursions.

My chief work in 1846 had been the survey of the parish of Llantwit-juxta-Neath. The agent of the Gnoll Estate had undertaken the valuation for the tithe commutation, and arranged with me to do the survey and make the map and the necessary copies. When all was finished and the valuation made, I was told that I must collect the payment from the various farmers in the parish, who would afterwards deduct it from their rent. This was a disagreeable business, as many of the farmers were very poor; some could not speak English, and could not be made to understand what it was all about; others positively refused to pay; and the separate amounts were often so small that it was not worth going to law about them, so that several were never paid at all, and others not for a year afterwards. This was one of the things that disgusted me with business, and made me more than ever disposed to give it all up if I could get anything else to do.

My brother John and I also obtained a little building and architectural work, and amongst other structures we designed and supervised the erection of the Mechanics' Institute at Neath. It was before the members of this institute that I made my first essay as a lecturer, having been persuaded to give a series of lectures on physics during two winters. I also gave a lecture describing a short visit I had made to Paris.[1]

During the two summers that I and my brother John lived at Neath we spent a good deal of our leisure time in wandering about this beautiful district, on my part in search of insects, while my brother always had his eyes open for any uncommon bird or reptile.

Though I have by no means a very wide acquaintance with the mountain districts of Britain, yet I know Wales pretty well; have visited the best parts of the lake district; in Scotland have been to Loch Lomond, Loch Katrine, and Loch Tay; have climbed Ben Lawers, and roamed through Glen Clova in search for rare plants;—but I cannot call to mind a single valley that in the same extent of country comprises so much beautiful and picturesque scenery, and so many interesting special features, as the Vale of Neath. The town itself is beautifully situated, with the fine wooded and rock-girt Drumau Mountain to the west, while immediately to the east are well-wooded heights crowned by Gnoll House, and to the south-east, three miles away, a high rounded hill, up which a chimney has been carried from the Cwm Avon copperworks in

[1] In 1895 I received a letter from Cardiff, from one of the workmen who attended the Neath Mechanics' Institution, asking if the author of "Island Life," the "Malay Archipelago," and other books is the same Mr. Alfred Wallace who taught in the evening science classes to the Neath Abbey artificers. He writes—" I have often had a desire to know, as I benefited more while in your class—if you are the same Mr. A. Wallace—than I ever was taught at school. I have often wished I knew how to thank you for the good I and others received from your teaching.—(Signed) MATTHEW JONES."

the valley beyond, the smoke from which gives the hill much the appearance of an active volcano. To the south-west the view extends down the valley to Swansea Bay, while to the north-east stretches the Vale of Neath itself, nearly straight for twelve miles, the river winding in a level fertile valley about a quarter to half a mile wide, bounded on each side by abrupt hills, whose lower slopes are finely wooded, and backed by mountains from fifteen hundred to eighteen hundred feet high. The view up this valley is delightful, its sides being varied with a few houses peeping out from the woods, abundance of lateral valleys and ravines, with here and there the glint of falling water, while its generally straight direction affords fine perspective effects, sometimes fading in the distance into a warm yellow haze, at others affording a view of the distant mountain ranges beyond.

At twelve miles from the town we come to the little village of Pont-nedd-fychan (the bridge of the little Neath river), where we enter upon a quite distinct type of scenery, dependent on our passing out of the South Wales coal basin, crossing the hard rock-belt of the millstone grit, succeeded by the picturesque crags of the mountain limestone, and then entering on the extensive formation of the Old Red Sandstone.

Within a mile of Pont-nedd-fychan is the Dinas rock, a tongue of mountain limestone jutting out across the millstone grit, and forming fine precipices, one of which was called the Bwa-maen or bow rock, from its being apparently bent double. Lower down there are also some curious waving lines of apparent stratification, but on a recent examination I am inclined to think that these are really glacial groovings caused by the ice coming down from Hirwain, right against these ravines and precipices, and being thus heaped up and obliged to flow away at right angles to its former course.

PORTH-YR-OGOF, VALE OF NEATH

But the most remarkable and interesting of the natural phenomena of the upper valley is Porth-yr-Ogof (the gatewy of the cavern), where the river Mellte runs for a quarter of a mile underground. The entrance is under a fine arch of limestone rock overhung with trees, as shown in the accompanying photograph. The outlet is more irregular and less lofty, and is also less easily accessible; but the valley just below has wooded banks, open glades, and fantastic rocks near the cave, forming one of the most charmingly picturesque spots imaginable.

I have already described one of the curious "standing stones" near the source of the Llia river, but there is a still more interesting example about a mile and a half north-west of Ystrad-fellte, where the old Roman road—the Sarn Helen—crosses over the ridge between the Nedd and the Llia valleys. This is a tall, narrow stone, roughly quadrilateral, on one of the faces of which there is a rudely inscribed Latin inscription, as seen in the photograph, and in a copy of the letters given. It reads as follows:——

DERVACI FILIUS JUSTI IC IACIT

meaning "[The body] of Dervacus the son of Justus lies here." It will be seen that the letters D, A, and I in Dervaci, and the T and I in Justi are inverted or reversed, probably indicating that the cutting was done by an illiterate workman, who placed them as most convenient when working on an erect stone. The stone itself is probably British, and was utilized as a memorial of a Roman soldier who died near the place.

One of our most memorable excursions was in June, 1846, when I and my brother spent the night in the water-cave. I wanted to go again to the top of the Beacons to see if I could find any rare beetles there, and also to show my brother the waterfalls and other beauties of the upper valley. Starting after an early breakfast we walked to Pont-nedd-fychan, and then turned up the western branch to the Rocking Stone, a large boulder of millstone-grit resting on a nearly level surface, but which by a succession of pushes with one hand can be made to rock considerably. It was here I obtained one of the most beautiful British beetles, *Trichius fasciatus*, the only time I ever captured it. We then went on to the Gladys and Einon Gam falls; then, turning back, followed up the river Nedd for some miles, crossed over to the cavern, and then on to Ystrad-fellte, where we had supper and spent the night, having walked leisurely about eighteen or twenty miles.

The next morning early we proceeded up the valley to the highest farm on the Dringarth, then struck across the mountain to the road from Hirwain to Brecon, which we followed to the bridge over the Taff, and then turned off towards the Beacons, the weather being perfect. It was a

MAEN MADOC.

delightful walk, on a gradual slope of fifteen hundred feet in a mile and a half, with a little steeper bit at the end, and the small overhanging cap of peat at the summit, as already described. I searched over it for beetles, which were, however, very scarce, and we then walked a little way down the southern slope to where a tiny spring trickles out—the highest source of the river Taff—and there, lying on the soft mountain turf, enjoyed our lunch and the distant view over valley and mountain to the faint haze of the Bristol Channel.

We took nearly the same route back, had a substantial tea at the little inn at Ystrad-fellte, and then, about seven o'clock, walked down, to the cave to prepare our quarters for the night. I think we had both of us at this time determined, if possible, to go abroad into more or less wild countries, and we wanted for once to try sleeping out-of-doors, with no shelter or bed but what nature provided.

Just inside the entrance of the cave there are slopes of water-worn rock and quantities of large pebbles and boulders, and here it was quite dry, while farther in, where there were patches of smaller stones and sand, it was much colder and quite damp, so our choice of a bed was limited to rock or boulders. We first chose a place for a fire, and then searched for sufficient dead or dry wood to last us the night. This took us a good while, and it was getting dusk before we lit our fire. We then sat down, enjoying the flicker

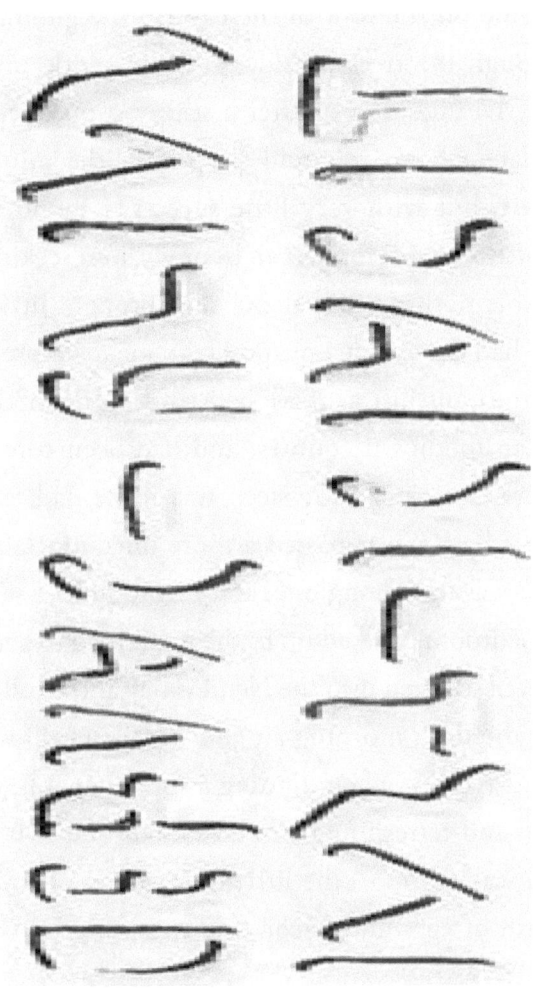

LATIN INSCRIPTIONS

of the flame on the roof of the cavern, the glimmer of the stars through the trees outside, and the gentle murmur of the little river beside us. After a scanty supper we tried to find a place where we could sleep with the minimum of discomfort, but with very little success. I found it almost impossible to lie still for half an hour without seeking a more comfortable position, but the change brought little relief. I think we had determined purposely to make no preparation, but to camp out just as if we had come accidentally to the place in an unknown country, and had been compelled to sleep there. But very little sleep was to be had, and while in health I have never passed a more uncomfortable night. Luckily it was not a long one, and before sunrise we left our gloomy bedroom, walked up to the main road to get into the sunshine, descended into the Nedd valley and strolled along, enjoying the fresh morning air and warm sun till we neared Pont-nedd-fychan, when, finding a suitable pool, we took a delightful and-refreshing bath, dried, our bodies in the sun, and then walked on to the little inn, where we enjoyed our ample dish of eggs and bacon, with tea, and brown bread-and-butter.

There is one subject on which I obtained conclusive evidence while living at Neath, which may here be briefly noticed. I have already described how at Leicester I became convinced of the genuineness of the phenomena of mesmerism, and was able thoroughly to test them myself. I also was able to make experiments which satisfied me of the truth of phrenology, and had read sufficient to enable me to understand its general principles. But during my early residence at Neath after my brother's death, I heard two lectures on the subject, and in both cases I had my character delineated with such accuracy as to render it certain that the positions of all the mental organs had been very precisely determined. It must be understood that the lecturers were both strangers, and that they each gave only a single lecture on their way to more important centres.

I will give the more detailed of the two delineations. It is as follows, only omitting a few words at the end which are of a purely personal nature:——

"*(a)* There is some delicacy in the nervous system, and consequent sensitiveness which unfits it for any very long-continued exertion; but this may be overcome by a strong will. There is some tendency to indigestion; this requires air and exercise.

(b) The power of fixing the attention is very good indeed, and there is very considerable perceptive power, so that this gentleman should learn easily and remember well, notwithstanding verbal memory is but moderate. Concentrativeness is the chief organ upon which all the memories depend, and this is undoubtedly large.

(c) He has some vanity, and more ambition. He may occasionally exhibit a want of self-confidence; but general opinion ascribes to him too much. In this, opinion is wrong: he knows that he has not enough; he may assume it, but it will sit ill.

(d) If Wit were larger he would be a good Mathematician; but without it, however clear and analytical the mind may be, it wants breadth and depth, and so I do not put down his mathematical talents as first-rate, although Number is good. The same must be said of his classical abilities—good, but not first-rate.

(e) He has some love for music from his Ideality, but I do not find a good ear, or sufficient time; he has, however, mechanical ability sufficient to produce enough of both, especially for the flute, if he so choose.

(f) As an artist, he would excel if his vision were perfect: he has every necessary faculty, even to Imitation.

(g) He is fond of argument, and not easily convinced; he would exhibit physical courage if called upon; and although he loves money—as who does not?—so far from there being any evidence of greediness, he is benevolent and liberal, but probably not extravagant. This part of his disposition is,

however, so evenly balanced that there is not likely to be much peculiarity.

(h) His domestic affections are his best.

Conscientiousness ought to be one more, but I do not see what will try it.

J. Q. RUMBALL.„

This delineation, and the other, which though not so complete agreed with it in nearly every particular, seem to me wonderfully accurate. That these two expositions of my character, the result of a very rapid examination of the form of my head by two perfect strangers, are so exact in many distinct points, demonstrates, I think, a large amount of truth—both in principle and in the details—of the method by which they were arrived at.

A. R. WALLACE. AGED 24.

(From a Daguerreotype.)

CHAPTER X

FOUR YEARS IN THE AMAZON VALLEY

DURING my residence at Neath I kept up some correspondence with H. W. Bates, chiefly on insect collecting. We exchanged specimens, and, I think in the summer of 1847, he came on a week's visit, which we spent chiefly in beetle-collecting and in discussing various matters, and it must have been at this time that we talked over a proposed collecting journey to the tropics, but had not then decided where to go. Mr. Bates' widow having kindly returned to me such of my letters as he had preserved, I find in them some references to the subjects in which I was then interested. I will, therefore, here give a few extracts from them.

I had recently read "Vestiges of the Natural History of Creation," and was much impressed by it, and I gave my views concerning it in several of my letters to Bates. In one letter I wrote, "It furnishes a subject for every observer of nature to attend to; every fact he observes will make either for or against it, and it thus serves both as an incitement to the collection of facts, and an object to which they can be applied when collected. Many eminent writers support the theory of the progressive development of animals and plants."

And in another letter—"I was much pleased to find that you so well appreciated Lyell. I first read Darwin's 'Journal' three or four years ago, and have lately re-read it. As the Journal of a scientific traveller, it is second only to Humboldt's 'Personal Narrative'—as a work of general interest, perhaps superior to it." My reference to Darwin's "Journal" and to Humboldt's "Personal Narrative" indicate, I believe, the two works to whose inspiration I owe my determination to visit the tropics as a collector.

In the last letter to Bates before our South American voyage I wrote—"I begin to feel rather dissatisfied with a mere local collection, little is to be learned by it. I should like to take some one family to study thoroughly, principally with a view to the theory of the origin of species. By that means I am strongly of opinion that some definite results might be arrived at." And at the very end of the letter I say—"There is a work published by the Ray Society I should much like to see, Oken's ' Elements of Physiophilosophy.' There is a review of it in the *Athenæum*. It contains some remarkable views on my favourite subject—the variations, arrangements, distribution, etc., of species."

These extracts from my early letters to Bates suffice to show that the great problem of the origin of species was already distinctly formulated in my mind; that I was not satisfied with the more or less vague solutions at that time offered; that I believed the conception of evolution through natural law so clearly formulated in the "Vestiges" to be, so far as it went, a true one; and that I firmly believed that a full and careful study of the facts of nature would ultimately lead to a solution of the mystery.

What decided our going to Para and the Amazon rather than to any other part of the tropics was the publication in 1847 of "A Voyage up the Amazon," by Mr. W. H. Edwards. This little book was so clearly and brightly written, described so well the beauty and the grandeur of tropical vegetation, and gave such a pleasing account of the people, while showing that expenses of living and of travelling were both very moderate, that Bates and myself at once agreed that this was the very place for us to go to if there was any chance of paying our expenses by the sale of our duplicate collections. We immediately communicated with Mr. Edward Doubleday, who had charge of the butterflies at the British Museum, for his advice upon the matter. He assured us that the whole of Northern Brazil was very little known, and that there was no doubt we could easily pay our expenses. Thus encouraged, we determined to go to Para, and began to make all the necessary arrangements.

After spending some time in London studying the insects in the British Museum, purchasing collecting apparatus and outfit, we set sail from Liverpool on April 20, 1848.

At this time there were very few steamships, and most

of the ocean trade was still carried on in sailing vessels. Ours was one of the smallest, being a barque of 192 tons named the *Mischief,* and said to be a very fast sailer. Bates and I were the only passengers.

We encountered violent storms in the Bay of Biscay, the waves flooded our decks, washed away part of our bulwarks, and very nearly swamped us altogether. All this time I was in my berth prostrate with sea-sickness. After the first six days the weather became fine, and the remainder of the voyage was very enjoyable.

We landed at Para on May 19, twenty-nine days after leaving Liverpool.

From this date till I landed at Deal in October, 1852, my adventures are narrated in my book "Travels on the Amazon and Rio Negro," and I will here, therefore, only give a very brief outline of my wanderings.

For the first four months Bates and I lived and collected together in and around Para, but on our return from an expedition which we had made up the Tocantins river, we agreed that it would be better for many reasons to travel and collect independently; one reason being that the country was so vast and so rich in birds and insects that much better results would be obtained if we each explored separate districts. We therefore separated, but we again met at Santarem and at the Barra. Afterwards Bates devoted himself to the Upper Amazon, while I ascended the Rio Negro and the unknown

Uaupés.

After spending about nine months collecting in the neighbourhood of Para, I went on an expedition up the river Guamá, on my return from which in July, 1849, my younger brother, Herbert, joined me with the idea of becoming a collector. We spent a month together in Para, and then went to Santarem, where we intended to stay for some time. Dr. Richard Spruce, the now well-known traveller and botanist, had come out in the same ship with my brother, and we much enjoyed his society during the short time he stayed with us here.

After spending three months at Santarem we went to the city of Barra at the mouth of the Rio Negro, but here we found the scarcity of insects and birds a great contrast to the abundance at Para, and after exploring the neighbouring country in all directions with poor results I determined to make a long expedition to the Upper Rio Negro, whilst my brother, who had finally decided to give up collecting, which had little interest for him, was to remain a few months at Barra, and then return to Para and England.

When he arrived at Para, and while waiting for his ship to sail, he was seized with yellow fever, and died in a few days.

The remainder of my South American travels consisted of two voyages up the Rio Negro. On the first I went beyond the boundaries of Brazil, and crossed by a road in the forest to one of the tributaries of the Orinoco. Returning thence I visited a village up a small branch of the Rio Negro, where there is an isolated rocky mountain, the haunt of the beautiful Cock of the Rock; afterwards going up the Uaupés as far as the second cataract at Juaurité. I then returned with my collections to Barra, having determined to go much farther up the Uaupés in order to obtain, if possible, the white umbrella bird which I had been positively assured was found there; and also in the hopes of finding some new and better collecting ground near the Andes. These journeys were made, but the second was cut short by delays and the wet season. My health also had suffered so much by a succession of fevers and dysentery that I did not consider it prudent to stay longer in the country.

Although during the last two journeys in the Rio Negro and Orinoco districts I had made rather large miscellaneous collections, and especially of articles of native workmanship, I never found any locality at all comparable with Para as a collecting ground. The numerous places I visited along more than a thousand miles of river, all alike had that poverty of insect and bird-life which characterized Barra itself, a poverty which is not altogether explicable. The enormous difficulties and delays of travel made it impossible to be at the right place at the right season; while the excessive wetness of the climate rendered the loss of the only month or two of fine weather irreparable for the whole year. The comparative scantiness of native population at all the towns of the Rio Negro, the small amount of cultivation, the scarcity of roads through the forest, and the want of any guide from the experience of previous collectors, combined to render my numerous journeys in this almost totally unknown region comparatively unproductive in birds and insects. As it happened (owing to Custom House formalities at Barra), the whole of my collections during the last two voyages were with me on the ship that was burnt, and were thus totally lost.

One letter I wrote from Guia on the Upper Rio Negro, three months after my arrival there, has been preserved, and from it I extract the following passage:——

"I have been spending a month with some Indians three days' journey up a narrow stream (called the Cobati River). From there we went half a day's journey through the forest to a rocky mountain where the celebrated 'Gallos de Serra' (Cocks of the Rock) breed. But we were very unfortunate, for though I had with me ten hunters and we remained nine days at the Serra, suffering many inconveniences (having only taken farinha and salt with us), I only got a dozen gallos, whereas I had expected in less time to have secured fifty. Insects there were none at all; and other good birds excessively rare.

"My canoe is now getting ready for a further journey

up to near the sources of the Rio Negro in Venezuela, where I have reason to believe I shall find insects more plentiful, and at least as many birds as here. On my return from there I shall take a voyage up the great River Uaupés, and another up the Isanna, not so much for my collections, which I do not expect to be very profitable there, but because I am so much interested in the country and the people that I am determined to see and know more of it and them than any other European traveller. If I do not get profit, I hope at least to get some credit as an industrious and persevering traveller."

Looking back over my four years' wanderings in the Amazon valley, there seem to me to be three great features which especially impressed me, and which fully equalled or even surpassed my expectations of them. The first was the virgin forest, everywhere grand, often beautiful and even sublime. Its wonderful variety with a more general uniformity never palled. Standing under one of its great buttressed trees—itself a marvel of nature—and looking carefully around, noting the various columnar trunks rising like lofty pillars, one soon perceives that hardly two of these are alike. The shape of the trunks, their colour and texture, the nature of their bark, their mode of branching and the character of the foliage far overhead, or of the fruits or flowers lying on the ground, have an individuality which shows that they are all distinct species differing from one another as our oak, elm, beech, ash, lime, and sycamore differ. This extraordinary variety of the species is a general though not universal characteristic of tropical forests, but seems to be nowhere so marked a feature as in the great forest regions which encircle the globe for a few degrees on each side of the equator. An equatorial forest is a kind of natural arboretum, where specimens of an immense number of species are brought together by nature. The western half of the island of Java affords an example of such a forest-region which has been well explored, botanically; and although almost all the fertile plains have been cleared for cultivation, and the forests cover only a small proportion of the country, the number of distinct species of forest-trees is said to be over fifteen hundred.

The second feature, that I can never think of without delight, is the wonderful variety and exquisite beauty of the butterflies and birds, a variety and charm which grow upon one month after month and year after year, as ever new and beautiful, strange and even mysterious, forms are continually met with. Even now I can hardly recall them without a thrill of admiration and wonder.

The third and most unexpected sensation of surprise and delight was my first meeting and living with man in a state of nature—with absolute uncontaminated savages! This was on the Uaupés River, and the surprise of it was that I did not in the least expect to be so surprised. I had already been two years in the country, always among Indians of many tribes; but these were all what are called tame Indians, they wore at least trousers and shirt; they had been (nominally) converted to Christianity, and were under the government of the nearest authorities; and all of them spoke either Portuguese or the common language, called "Lingoa-Geral."

But these true wild Indians of the Uaupés were at once seen to be something totally different. They had nothing that we call clothes; they had peculiar ornaments, tribal marks, etc.; they all carried weapons or tools of their own manufacture; they were living, for the most part, in large houses, many families together, quite unlike the huts of the tame Indians; but, more than all, their whole aspect and manner were different—they were all going about their own work or pleasure which had nothing to do with white men or their ways; they walked with the free step of the independent forest-dweller, and, except the few that were known to my companion, paid no attention whatever to us, mere strangers of an alien race. In every detail they were original and self-sustaining as are the wild animals of the forests, absolutely independent of civilization, and who could and did live their own lives in their own way, as they had done for countless generations before America was discovered. I could not have believed that there would be so much difference in the aspect of the same people in their native state and when living under European supervision.

The true denizen of the Amazonian forests, like the forest itself, is unique and not to be forgotten.

My voyage home was rather adventurous. I will therefore print here a letter written to Dr. Spruce, which has not before been made public, and which describes it.

"Brig *Fordeson*, N. Lat. 49° 30′, W. Long. 20°.

Sunday, September 19, 1852.

MY DEAR FRIEND,

Having now some prospect of being home in a week or ten days, I will commence giving you an account of the peculiar circumstances which have already kept me at sea seventy days on a voyage which took us only twenty-nine days on our passage out. I hope you have received the letter sent you from Para, dated July 9 or 10, in which I informed you that I had taken my passage in a vessel bound for London, which was to sail in a few days. On Monday, July 12, I went on board with all my cargo, and some articles purchased or collected on my way down, with the remnant (about twenty) of my live stock.[1] After being at sea about a week I had a slight attack of fever, and at first thought I had got the yellow fever after all. However, a little calomel set me right in a few days, but I remained rather weak, and spent most of my time reading in the cabin, which was very comfortable. On Friday, August 6, we were in N. Lat. 30° 30′, W. Long. 52°, when, about nine in the morning, just after breakfast, Captain Turner, who was half-owner of the vessel, came into the cabin, and said, ' I'm afraid the ship's on fire. Come and see what you think of it.' Going on deck I found a thick smoke coming out of the forecastle, which we both thought more like the steam from heating vegetable matter than the smoke from a fire. The fore hatchway was immediately opened to try and ascertain the origin of the smoke, and a quantity of cargo was thrown out, but the smoke continuing without any perceptible increase, we went to the after hatchway, and after throwing out a quantity of piassaba, with which the upper part of the hold was filled, the smoke became so dense that the men could not stay in it. Most of them were then set to work throwing in buckets of water, and the rest proceeded to the cabin and opened the lazaretto or store-place beneath its floor, and found smoke issuing from the bulkhead separating it from the hold, which extended halfway under the fore part of the cabin. Attempts were then made to break down this bulkhead, but it resisted all efforts, the smoke being so suffocating as to prevent any one stopping in it more than a minute at a time. A hole was then cut in the cabin floor, and while the carpenter was doing this, the rest of the crew were employed getting out the boats, the captain looked after his chronometer, sextant, books, charts and compasses, and I got up a small tin box containing a few shirts, and put in it my drawings of fishes and palms, which were luckily at hand; also my watch and a purse with a few sovereigns. Most of my clothes were scattered about the cabin, and in the dense suffocating smoke it was impossible to look about after them. There were two boats, the long-boat and the captain's gig, and it took a good deal of time to get the merest necessaries collected and put into them, and to lower them into the water. Two casks of biscuit and a cask of water were got in, a lot of raw pork and some ham, a few tins of preserved meats and vegetables, and some wine. Then there were corks to stop the holes in the boats, oars, masts, sails, and rudders to be looked up, spare spars, cordage, twine, canvas, needles, carpenter's tools, nails, etc. The crew brought up their bags of clothes, and all were bundled indiscriminately into the boats, which, having been so long in the sun, were very leaky, and soon became half full of water, so that two men in each of them had to be constantly baling out the water with buckets. Blankets, rugs, pillows, and clothes were all soaked, and the boats seemed overloaded, though there was really very little weight in them. All being now prepared, the crew were again employed pouring water in the cabin and hatchway.

¹ These consisted of numerous parrots and parrakeets, and several uncommon monkeys, a forest wild-dog, etc.

The cargo of the ship consisted of rubber, cocoa, anatto, balsam-capivi, and piassaba. The balsam was in small casks, twenty stowed in sand, and twenty small kegs in rice-chaff, immediately beneath the cabin floor, where the fire seemed to be. For some time we had heard this bubbling and hissing as if boiling furiously, the heat in the cabin was very great, flame soon broke into the berths and through the cabin floor, and in a few minutes more blazed up through the skylight on deck. All hands were at once ordered into the boats, which were astern of the ship. It was now about twelve o'clock, only three hours from the time the smoke was first discovered. I had to let myself down into the boat by a rope, and being rather weak it slipped through my hands and took the skin off all my fingers, and finding the boat still half full of water I set to baling, which made my hands smart very painfully. We lay near the ship all the afternoon, watching the progress of the flames, which soon covered the hinder part of the vessel, and rushed up the shrouds and sails in a most magnificent conflagration. Soon afterwards, by the rolling of the ship, the masts broke off and fell overboard, the decks soon burnt away, the ironwork at the sides became red-hot, and last of all the bowsprit, being burnt at the base, fell also. No one had thought of being hungry till darkness came on, when we had a meal of biscuit and raw ham, and then disposed ourselves as well as we could for the night, which, you may be sure, was by no means a pleasant one. Our boats continued very leaky, and we could not cease an instant from baling; there was a considerable swell, though the day had been remarkably fine, and there were constantly floating around us pieces of the burnt wreck, masts, etc., which might have stove in our boats had we not kept a constant look-out to keep clear of them. We remained near the ship all night, in order that we might have the benefit of its flames attracting any vessel that might pass within sight of it.

It now presented a magnificent and awful sight as it rolled over, looking like a huge caldron of fire, the whole cargo of rubber, etc., forming a liquid burning mass at the bottom. In the morning our little masts and sails were got up, and we bade adieu to the *Helen*, now burnt down to the water's edge, and proceeded with a light east wind towards the Bermudas, the nearest land, but which were more than seven hundred miles from us. As we were nearly in the track of West Indian vessels, we expected to fall in with some ship in a few days.

I cannot attempt to describe my feelings and thoughts during these events. I was surprised to find myself very cool and collected. I hardly thought it possible we should escape, and I remember thinking it almost foolish to save my watch and the little money I had at hand. However, after being in the boats some days I began to have more hope, and regretted not having saved some new shoes, cloth coat and trousers, hat, etc., which I might have done with a little trouble. My collections, however, were in the hold, and were irretrievably lost. And now I began to think that almost all the reward of my four years of privation and danger was lost. What I had hitherto sent home had little more than paid my expenses, and what I had with me in the *Helen* I estimated would have realized about £500. But even all this might have gone with little regret had not by far the richest part of my own private collection gone also. All my private collection of insects and birds since I left Para was with me, and comprised hundreds of new and beautiful species, which would have rendered (I had fondly hoped) my cabinet, as far as regards American species, one of the finest in Europe. Fancy your regrets had you lost all your Pyrenean mosses on your voyage home, or should you now lose all your South American collection, and you will have some idea of what I suffer. But besides this, I have lost a number of sketches, drawings, notes, and observations on natural history, besides the three most interesting years of my journal, the whole of which, unlike any pecuniary loss, can never be replaced; so you will see that I have some need of philosophic resignation to bear my fate with patience and equanimity.

Day after day we continued in the boats. The winds changed, blowing dead from the point to which we wanted to go. We were scorched by the sun, my hands, nose, and ears being completely skinned, and were drenched continually by the seas or spray. We were therefore almost constantly wet, and had no comfort and little sleep at night. Our meals consisted of raw pork and biscuit, with a little preserved meat or carrots once a day, which was a great luxury, and a short allowance of water, which left us as thirsty as before directly after we had drunk it. Ten days and ten nights we spent in this manner. We were still two hundred miles from Bermuda, when in the afternoon a vessel was seen, and by eight in the evening we were on board her, much rejoiced to have escaped a death on the wide ocean, whence none would have come to tell the tale. The ship was the *Fordeson*, bound for London, and proves to be one of the slowest old ships going. With a favourable wind and all sail set, she seldom does more than five knots, her average being two or three, so that we have had a most tedious time of it, and even now cannot calculate with any certainty as to when we shall arrive. Besides this, she was rather short of provisions, and as our arrival exactly doubled her crew, we were all obliged to be put on strict allowance of bread, meat, and water. A little ham and butter of the captain's were soon used up, and we have been now for some time on the poorest of fare. We have no suet, butter, or raisins with which to make ' duff,' or even molasses, and barely enough sugar to sweeten our tea or coffee, which we take with dry, coarse biscuit, and for dinner, beef or pork of the very worst quality I have ever eaten or even imagined to exist. This, repeated day after day without any variation, beats even Rio Negro fare, rough though it often was. About a week after we were picked up we spoke and boarded an outward-bound ship, and got from her some biscuits, a few potatoes, and some salt cod, which were a great improvement, but did not last long. We have also occasionally caught some dolphin and a few fish resembling the acarrás of the Rio Negro; but for some time now we had seen none, so that I am looking forward to the 'flesh-pots of Egypt' with as much pleasure as when we were luxuriating daily on farinha and 'fiel amigo.'[1] While we were in the boats we had generally fine weather, though with a few days and nights squally and with a heavy sea, which made me often tremble for our safety, as we heeled over till the water poured in over the boat's side. We had almost despaired of seeing any vessel, our circle of vision being so limited; but we had great hopes of reaching Bermuda, though it is doubtful if we should have done so, the neighbourhood of those islands being noted for sudden squalls and hurricanes, and it was the time of year when the hurricanes most frequently occur. Having never seen a great gale or storm at sea, I had some desire to witness the phenomenon, and have now been completely gratified. The first we had about a fortnight ago. In the morning there was a strong breeze and the barometer had fallen nearly half an inch during the night and continued sinking, so the captain commenced taking in sail, and while getting in the royals and studding-sails, the wind increased so as to split the mainsail, fore-topsail, fore-trysail, and jib, and it was some hours before they could be got off her, and the main-topsail and fore-sail double reefed. We then went flying along, the whole ocean a mass of boiling foam, the crests of the waves being carried in spray over our decks. The sea did not get up immediately, but by night it was very rough, the ship plunging and rolling most fearfully, the sea pouring in a deluge over the top of her bulwarks, and sometimes up over the cabin skylight. The next morning the wind abated, but the ship, which is a very old one, took in a deal of water, and the pumps were kept going nearly the whole day to keep her dry. During this gale the wind went completely round the compass, and then settled nearly due east, where it pertinaciously continued for twelve days, keeping us tacking about, and making less than forty miles a day against it. Three days ago we had another gale, more severe than the former one—a regular equinoctial, which lasted two entire days and nights, and split one of the newest and strongest sails on the ship. The rolling and plunging were fearful, the bowsprit going completely under water, and the ship being very heavily laden with mahogany, fustic, and other heavy woods from Cuba, strained and creaked tremendously, and leaked to that extent that the pumps were obliged to be kept constantly going, and their continued click-clack, click-clack all through the night was a most disagreeable and nervous sound. One day no fire could be made owing to the sea breaking continually into the galley, so we had to eat a biscuit for our dinner; and not a moment's rest was to be had, as we were obliged to be constantly holding on, whether standing, sitting, or lying, to prevent being pitched about by the violent plunges and lurches of the vessel. The gale, however, has now happily passed, and we have a fine breeze from the northwest, which is taking us along six or seven knots— quicker than we have ever gone yet. Among our other disagreeables here we have no fresh water to spare for washing, and as I only saved a couple of shirts, they are in a state of most uncomfortable dirtiness, but I console myself with the thoughts of a glorious warm bath when I get on shore.

[1] This was the name given by our kind host, Señor Henrique, at Barra to dried pirarucú, meaning "faithful friend," always at hand when other food failed.

* * * * *

October I. Oh, glorious day! Here we are on shore at Deal, where the ship is at anchor. Such a dinner, with our two captains! Oh, beef-steaks and damson tart, a paradise for hungry sinners.

* * * * *

October 5, London. Here I am laid up with swelled ankles, my legs not being able to stand work after such a long rest in the ship. I cannot write now at any length—I have too much to think about. We had a narrow escape in the Channel. Many vessels were lost in a storm on the night of September

29, but we escaped. The old 'Iron Duke' is dead. The Crystal Palace is being pulled down, and is being rebuilt on a larger and improved plan by a company. Loddige's collection of plants has been bought entire to stock it, and they think by heating it in the centre to get a gradation of climates, so as to be able to have the plants of different countries, tropical or temperate, in one undivided building. This is Paxton's plan.

How I begin to envy you in that glorious country where 'the sun shines for ever unchangeably bright,' where farinha abounds, and of bananas and plantains there is no lack! Fifty times since I left Para have I vowed, if I once reached England, never to trust myself more on the ocean. But good resolutions soon fade, and I am already only doubtful whether the Andes or the Philippines are to be the scene of my next wanderings. However, for six months I am a fixture here in London, as I am determined to make up for lost time by enjoying myself as much as possible for awhile. I am fortunate in having about £200 insured by Mr. Stevens' foresight, so I must be contented, though it is very hard to have nothing to show of what I took so much pains to procure.

I trust you are well and successful. Kind remembrances to everybody, everywhere, and particularly to the respectable Senhor Joaõ de Lima of Saõ Joachim.

Your very sincere friend,
ALFRED R. WALLACE."

Some of the most alarming incidents, to a landsman, are not mentioned either in this letter or in my published "Narrative." The captain had given the only berths in the cabin to Captain Turner and myself, he sleeping on a sofa in fine weather, and on a mattress on the floor of the cabin when rough. On the worst night of the storm I saw him, to my surprise, bring down an axe and lay it beside him, and on asking what it was for, he replied, "To cut away the masts in case we capsize in the night." In the middle of the night a great sea smashed our skylight and poured in a deluge of water, soaking the poor captain, and then slushing from side to side with every roll of the ship. Now, I thought, our time is come; and I expected to see the captain rush up on deck with his axe. But he only swore a good deal, sought out a dry coat and blanket, and then lay down on the sofa as if nothing had happened. So I was a little reassured.

Not less alarming was the circumstance of the crew coming aft in a body to say that the forecastle was uninhabitable as it was constantly wet, and several of them brought handfuls of wet rotten wood which they could pull out in many places. This happened soon after the first gale began; so the two captains and I went to look, and we saw sprays and squirts of water coming in at the joints in numerous places, soaking almost all the men's berths, while here and there we could see the places where they had pulled out rotten wood with their fingers. The captain then had the sail-room amid-ships cleared out for the men to sleep in for the rest of the voyage.

One day in the height of the storm, when we were being flooded with spray and enormous waves were coming up behind us, Captain Turner and I were sitting on the poop in the driest place we could find, and, as a bigger wave than usual rolled under us and dashed over our sides, he said quietly to me, "If we are pooped by one of those waves we shall go to the bottom;" then added, "We were not very safe in our two small boats, but I had rather be back in them where we were picked up than in this rotten old tub." It is, therefore, I think, quite evident that we *did* have a very narrow escape. Yet this unseaworthy old ship, which ought to have been condemned years before, had actually taken Government stores out to Halifax, had there been patched up, and sent to Cuba for a cargo of heavy timber, which we were bringing home.

A. R. WALLACE. AGED 30.

CHAPTER XI

LONDON, AND VOYAGE TO THE EAST

ON reaching London in the condition described in my letter to Dr. Spruce, and my only clothing a suit of the thinnest calico, I was met by my kind friend and agent, Mr. Samuel Stevens, who took me first to the nearest ready-made clothes shop, where I got a warm suit, then to his own tailor, where I was measured for what clothes I required, and afterwards to a haberdasher's to get a small stock of other necessaries. Having at that time no relatives in London, his mother, with whom he lived in the south of London— I think in Kennington—had invited me to stay with her. Here I lived most comfortably for a week, enjoying the excellent food and delicacies Mrs. Stevens provided for me, which quickly restored me to my usual health and vigour.

As I wished to be with my sister and mother during my stay in England, I took a house then vacant in Upper Albany Street (No.44), so that we might all live together. While it was getting ready I took lodgings next door, as the situation was convenient, being close to the Regent's Park and Zoological Gardens, and also near the Society's offices in Hanover Square, and within easy access to Mr. Stevens's office close to the old British Museum. At Christmas we were all comfortably settled, and I was able to begin the work which I had determined to do before again leaving England.

In the small tin box which I had saved from the wreck I fortunately had a set of careful pencil drawings of all the different species of palms I had met with, together with notes as to their distribution and uses. I had also a large number of drawings of fish, as already stated, carefully made to scale, with notes of their colours, their dentition, and their fin-rays, scales, etc. I had also a folio Portuguese note-book containing my diary while on the Rio Negro, and some notes and observations made for a map of that river and the Uaupés. With these scanty materials, helped by the letters I had sent home, I now set to work to write an account of my travels, as well as a few scientific papers for which I had materials in the portion of my collections made in Para, Santarem, and the Lower Rio Negro. These I had sent off before leaving Barra on my first voyage up the Rio Negro, and they had arrived home safely; but I had reserved all my private collections for comparison with future discoveries, and though I left these to be sent home before starting on my second voyage up the Rio Negro, they were never despatched, owing to the Custom House authorities at Barra insisting on seeing the contents before allowing them to go away. I therefore found them at Barra on my way home, and they were all lost with the ship.

As my collections had now made my name well known to the authorities of the Zoological and Entomological Societies, I received a ticket from the former, giving me admission to their gardens while I remained in England, and I was a welcome visitor at the scientific meetings of both societies, which I attended very regularly, and thus made the acquaintance of most of the London zoologists and entomologists. I also went frequently to examine the insect and bird collections in the British Museum (then in Great Russell Street), and also to the Linnean Society, and to the Kew Herbarium to consult works on botany, in order to name my palms.

After discussing the matter with some of my friends, I determined to publish, at my own expense, a small, popular volume on the "Palms of the Amazon and Rio Negro," with an account of their uses and distribution, and figures of all the species from my sketches and specimens of fruits. I arranged with Mr. Walter Fitch of Kew, the first botanical artist of the day, to draw them on stone, adding some artistic touches to give them life and variety, and in a few cases some botanical details from species living in the gardens. In one of the drawings a large native house on the Uaupés is introduced, with some figures which, I am sorry to say, are as unlike the natives as are the inhabitants of a London slum. I arranged with Mr. Van Voorst to publish this small volume, and it was not thought advisable to print more than 250 copies, the sale of which just covered all expenses.

At the same time I was preparing my "Travels on the Amazon and Rio Negro" from the scanty materials I had saved, supplemented by the letters I had written home. I arranged with Mr. Lovel Reeve for its publication on an agreement for "half profits." Only 750 copies were printed, and when I returned home from the East in 1862, about 250 copies were still unsold, and there were consequently no profits to divide. We agreed, however, to share the remaining copies, and my portion was disposed of by my new publisher, Messrs. Macmillan & Co., and brought me in a few pounds. I had brought with me vocabularies of about a hundred common words in ten different Indian languages, and as the great philologist at that time was the late Dr. R. G. Latham, I obtained an introduction to him, and he kindly offered to write some "Remarks" upon the vocabularies, and these are published in the first edition of my "Travels."

Dr. Latham was at this time engaged in fitting up groups of figures to illustrate the family life and habits of the various races of mankind at the new Crystal Palace at Sydenham, then just completed, and he asked me to meet him there and see whether any alterations were required in a group of natives, I think, of Guiana.

I found Dr. Latham among a number of workmen in white aprons, several life-size clay models of Indians, and a number of their ornaments, weapons, and utensils. The head modellers were Italians, and Dr. Latham told me he could get no Englishmen to do the work, and that these Italians, although clever modellers of the human figure in any required attitude, had all been trained in the schools of classical sculpture, and were unable to get away from this training. The result was very curious, and often even ludicrous, a brown Indian man or girl being given the attitudes and expressions of an Apollo or a Hercules, a Venus or a Minerva. In those days there were no photographs, and the ethnologist had to trust to paintings or drawings, usually exaggerated or taken from individuals of exceptional beauty or ugliness. Under my suggestion alterations were made both in the features and pose of one or two of the figures just completed, so as to give them a little more of the Indian character, and serve as a guide in modelling others, in which the same type of physiognomy was to be preserved. I went several times during the work on the groups of South American origin, but though when completed, with the real ornaments, clothing, weapons, and domestic implements, the groups were fairly characteristic and life-like, yet there remained occasionally details of attitude or expression which suggested classic Greece or Italy rather than the South American savage.

These ethnological figures, although instructive to the student, were never very popular, and soon became the subject of contempt and ridicule. One reason of this was their arrangement in the open, quite close to the passing visitor, with nothing to isolate them from altogether incongruous surroundings. Another was, that they were not carefully attended to, and when I saw them after my return from the East, they had a shabby and dilapidated appearance, and the figures themselves were more or less dusty, which had a most ludicrous effect in what were intended to represent living men and women, being so utterly unlike the clear, glossy, living skins of all savage peoples. To be successful and life-like, such groups should be each completely isolated in a deep recess, with three sides representing houses or huts, or the forest, or river-bank, while the open front should be enclosed by a single sheet of plate-glass, and the group should be seen at a distance of at least ten or fifteen feet. In this way, with a carefully arranged illumination from above and an artistic colouring of the figures and accessories, each group might be made to appear as life-like as some of the best figures at Madame Tussaud's, or as the grand interiors of cathedrals, which were then exhibited at the Diorama. In the museum of the future, such groups will find their place in due succession to the groups illustrating the life histories of the other mammalia; but ample space and a very careful attention to details must be given in order to ensure a successful and attractive representation.

It was at this time that I first saw Huxley. At one of the evening meetings of the Zoological Society (in December,

1852) he gave an account of some Echinococci found in the liver of a zebra which died in the gardens. He did not read the paper, but, with the help of diagrams and sketches on the blackboard, showed us clearly its main points of structure, its mode of development, and the strange transformations it underwent when the parent worm migrated from the intestine to other parts of the body of the animal. I was particularly struck with his wonderful power of making a difficult and rather complex subject perfectly intelligible and extremely interesting to persons who, like myself, were absolutely ignorant of the whole group. Although he was two years younger than myself, Huxley had already made a considerable reputation as a comparative anatomist, was a Fellow of the Royal Society, and a few months later was appointed Professor of Natural History and Palæontology at the Royal School of Mines. I was amazed, too, at his complete mastery of the subject, and his great amount of technical knowledge of a kind to which I have never given any attention, the structure and development of the lower forms of animal life. From that time I always looked up to Huxley as being immeasurably superior to myself in scientific knowledge, and supposed him to be much older than I was. Many years afterwards I was surprised to find that he was really younger.

During my constant attendance at the meetings of the Zoological and Entomological Societies, and visits to the insect and bird departments of the British Museum, I had obtained sufficient information to satisfy me that the very finest field for an exploring and collecting naturalist was to be found in the great Malayan Archipelago, of which just sufficient was known to prove its wonderful richness, while no part of it, with the one exception of the island of Java, has been well explored as regards its natural history. Sir James Brooke had recently become Rajah of Sarawak, while the numerous Dutch settlements in Celebes and the Moluccas offered great facilities for a traveller. So far as known also, the country was generally healthy, and I determined that it would be much better for me to go to such a new country than to return to the Amazon, where Bates had already been successfully collecting for five years, and where I knew there was a good bird-collector who had been long at work in the upper part of the river towards the Andes.

As the journey to the East was an expensive one, I was advised to try to get a free passage in some Government ship. Through my paper on the Rio Negro, I had made the acquaintance of Sir Roderick Murchison, then President of the Royal Geographical Society, and one of the most accessible and kindly of men of science. On calling upon him and stating my wishes, he at once agreed to make an application on my behalf for a passage to some Malayan port, and as he was personally known to many members of the Government and had great influence with them, a passage was promised me on the first ship going to those seas. This was, I think, near the end of the year 1853, when I had published my two books, and had spent much of my spare time at the British Museum, examining the collections, and making notes and sketches of the rarer and more valuable species of birds, butterflies, and beetles of the various Malay islands.

It was, I believe, in the latter part of January, 1854, that I received a notification from the Government that a passage had been granted me to Singapore in the brig *Frolic*, shortly sailing for that port, and that I was to communicate with the captain— Commander Nolloth—as to when I should go on board. I think it was about the middle of February that I went to Portsmouth with all necessaries for the voyage, my heavy baggage having been sent off by a merchant ship some time previously. The *Frolic* was anchored at Spithead and I went on board at once. I was kindly received by the captain and officers and found my quarters very comfortable.

Sailing orders were expected every day, as the ship was quite ready, with the stores she was taking out to the East all on board; but day after day and week after week passed, signals were exchanged with the admiral, but we seemed no nearer sailing than when I came on board. Then at last, one day the captain informed me that he had received fresh orders to carry stores to the Crimea, where the great war with Russia was about to commence. He said I had better leave the next morning, and that no doubt the Government would provide me a passage in some other vessel. So I bade farewell to him and his officers, none of whom I ever met again.

On returning to London, I at once called on Sir Roderick Murchison, and through his representations I received in a few days a first-class ticket to Singapore by the next Peninsular and Oriental steamer, which sailed in about a week, so that I did not lose much time. The voyage was a very interesting one, as we stopped a few hours at Gibraltar, passed within sight of the grand Sierra Nevada of Spain, stayed a day at Malta, where the town and the tombs of the knights were inspected, and then on to Alexandria. Having by me a long letter I wrote to my schoolfellow, Mr. George Silk, I will here quote from it a few of the impressions of my journey as they appeared to me at the time they occurred:—

"Steamer *Bengal*, Red Sea, March 26.

Of all the eventful days in my life (so far), my first in Alexandria was (in some respects) the most exciting. Imagine my feelings when, coming out of the hotel (to which we had been conveyed in an omnibus) with the intention of taking a quiet stroll through the city, I suddenly found myself in the midst of a vast crowd of donkeys and their drivers, all thoroughly determined to appropriate my person to their own use and interest, without in the least consulting my inclinations. In vain with rapid strides and waving arms I endeavoured to clear a way and move forward, arms and legs were seized upon, and even the Christian coat-tails were not sacred from the profane hands of the Mahometan crowd. One would hold together two donkeys by their tails whilst I was struggling between them, and another, forcing their heads together, hoped to compel me to mount one or both of them. One fellow, more impudent than the rest, I laid flat upon the ground, and, sending the little donkey staggering after him, I escaped for a moment midst hideous yells and most unearthly cries. I now beckoned to a fellow more sensible-looking than the rest, and told him that I wished to walk, and would take him as a guide, and now hoped that I might be left at peace. But vain thought! I was in the hands of the Philistines, who, getting me up against a wall, formed around me an impenetrable phalanx of men and brutes, thoroughly determined that I should only escape from the spot upon the four legs of a donkey. So, bethinking myself that donkey-riding was a national institution of venerable antiquity, and seeing a fat Yankee already mounted, being like myself, hopeless of any other means of escape, I seized upon a bridle in hopes that I should then be left by the remainder of the crowd. But seeing that I was at last going to ride, each one was determined that he alone should profit by the transaction, and a dozen animals were forced suddenly upon me, and a dozen pair of hands tried to lift me upon their respective beasts. But now my patience was exhausted, so, keeping firm hold of the bridle I had first taken with one hand, I hit right and left with the other, and calling upon my guide to do the same, we succeeded in clearing a little space around us. Now, then, behold your long-legged friend mounted upon a jackass in the streets of Alexandria; a boy behind, holding by his tail and whipping him up; Charles, who had been lost sight of in the crowd, upon another; and my guide upon a third; and off we go among a crowd of Jews and Greeks, Turks and Arabs, and veiled women and yelling donkey-boys, to see the city. We saw the bazaars, and the slave market (where I was again nearly pulled to pieces for 'backsheesh'), the mosques with their graceful minarets, and then the pasha's new palace, the interior of which is most gorgeous. We passed lots of Turkish soldiers, walking in comfortable irregularity; and after the consciousness of being dreadful guys for two crowded hours, returned to the hotel, whence we are to start for the canal boats. You may think this little narrative is exaggerated, but it is not so. The pertinacity, vigour, and screams of the Alexandrian donkey-drivers *cannot* be exaggerated. On our way to the boats we passed Pompey's Pillar; for a day we were rowed in small boats on a canal, then on the Nile in barges, with a panorama of mud villages, palm-trees, camels, and irrigating wheels turned by buffaloes,—a perfectly flat country, beautifully green with crops of corn and lentils; endless boats with immense triangular sails. Then the Pyramids came in sight, looking huge and solemn; then a handsome castellated bridge for the Alexandria and Cairo railway; and then Cairo—Grand Cairo! the city of romance, which we reached just before sunset. We took a guide and walked in the city, very picturesque and very dirty. Then to a quiet English hotel, where a Mussulman waiter, rejoicing in the name of Alibaba, gave us a splendid tea, brown bread and fresh butter. One or two French and English travellers were the only guests, and I could hardly realize my situation. I longed for you to enjoy it with me. Thackeray's 'First Day in the East' is admirable. Read it again, and you will understand just how I think and feel.

Next morning at seven we started for Suez in small four-horsed two-wheeled omnibuses, carrying six passengers each. Horses were changed every five miles, and we had a meal every three hours at very comfortable stations. The desert is undulating, mostly covered with a coarse, volcanic-looking gravel. The road is excellent. The skeletons of camels— hundreds of them—lay all along the road; vultures, sand-grouse, and sand-larks were occasionally seen. We frequently saw the mirage, like distant trees and water. Near the middle station the pasha has a hunting-lodge—a perfect palace. The Indian and Australian mails, about six hundred boxes, as well as all the parcels, goods, and passengers' luggage, were brought by endless trains of camels, which we passed on the way. At the eating-places I took a little stroll, gathering some of the curious highly odoriferous plants that grew here and there in the hollows, which I dried in my pocketbooks, and I also found a few land-shells. We enjoyed the ride exceedingly, and reached Suez about midnight. It is a miserable little town, and the bazaar is small, dark, and dirty. There is said to be no water within ten miles. The next afternoon we went on board our ship, a splendid vessel with large and comfortable cabins, and everything very superior to the *Euxine*. Adieu."

I have given this description of my journey from Alexandria to Suez, over the route established by Lieutenant Waghorn, which was superseded a few years later by the railway, and afterwards by the canal, because few persons now living will remember it, or know that it ever existed. Of the rest of our journey I have no record. We stayed a day at desolate, volcanic Aden, and thence sailed across to Galle, with its groves of cocoa-nut palms, and crowds of natives offering for sale the precious stones of the country; thence across to Pulo Penang, with its picturesque mountain, its spice-trees, and its waterfall, and on down the Straits of Malacca, with its richly wooded shores, to our destination, Singapore, where I was to begin the eight years of wandering throughout the Malay Archipelago, which constituted the central and controlling incident of my life.

CHAPTER XII

THE MALAY ARCHIPELAGO

(1854–1858)

MY wanderings and adventures in the far East have been recorded in my book "The Malay Archipelago." I will therefore give here but a brief outline with a few extracts from my letters and references to subjects of special interest.

I remained at Singapore for several months collecting insects and birds in the forests around. In a letter home I give a short account of my daily life at this time:— "I will tell you how my day is now occupied. Get up at half-past five, bath, and coffee. Sit down to arrange and put away my insects of the day before, and set them in a safe place to dry. Charles mends our insect-nets, fills our pin-cushions, and gets ready for the day. Breakfast at eight; out to the jungle at nine. We have to walk about a quarter mile up a steep hill to reach it, and arrive dripping with perspiration. Then we wander about in the delightful shade along paths made by the Chinese wood-cutters till two or three in the afternoon, generally returning with fifty or sixty beetles, some very rare or beautiful, and perhaps a few butterflies. Change clothes and sit down to kill and pin insects, Charles doing the flies, wasps, and bugs; I do not trust him yet with beetles. Dinner at four, then at work again till six: coffee. Then read or talk, or, if insects very numerous, work again till eight or nine. Then to bed."

Charles was a boy of sixteen whom I had brought with me from London as he wished to become a collector. He remained with me about a year and a half and eventually got employment on some of the plantations near Singapore.

I next made an expedition to Malacca which I describe in one of my letters as follows:— "I have now just returned to Singapore after two months' hard work. At Malacca I had a strong touch of fever, with the old 'Rio Negro' symptoms, but the Government doctor made me take large doses of quinine every day for a week, and so killed it, and in less than a fortnight I was quite well, and off to the jungle again. I never took half enough quinine in America to cure me.

"Malacca is a pretty place. Insects are not very abundant there, still, by perseverance, I got a good number, and many rare ones. Of birds, too, I made a good collection. I went to the celebrated Mount Ophir, and ascended to the top, sleeping under a rock. The walk there was hard work, thirty miles through jungle in a succession of mud-holes, and swarming with leeches, which crawled all over us, and sucked when and where they pleased. We lived a week at the foot of the mountain, in a little hut built by our men, near a beautiful rocky stream. I got some fine new butterflies there, and hundreds of other new or rare insects. Huge centipedes and scorpions, some nearly a foot long, were common, but we none of us got bitten or stung. We only had rice, and a little fish and tea, but came home quite well. The mountain is over four thousand feet high. Near the top are beautiful ferns and pitcher-plants, of which I made a small collection. Elephants and rhinoceroses, as well as tigers, are abundant there, but we had our usual bad luck in seeing only their tracks. On returning to Malacca I found the accumulation of two or three posts—a dozen letters, and about fifty newspapers. hellip; I am glad to be safe in Singapore with my collections, as from here they can be insured. I have now a fortnight's work to arrange, examine, and pack them, and four months hence there will be work for Mr. Stevens.[1]

Sir James Brooke is here. I have called on him. He received me most cordially, and offered me every assistance at Sarawak. I shall go there next, as I shall have pleasant society at Sarawak, and shall get on in Malay, which is very easy; but I have had no practice yet, though I can ask for most common things."

I reached Sarawak early in November, and remained in Borneo fourteen months, seeing a good deal of the country. The first four months was the wet season, during which I made journeys up and down the Sarawak river, but obtained very scanty collections. In March I went to the Sadong river, where coal mines were being opened by an English mining engineer, Mr. Coulson, a Yorkshireman, and I stayed there nearly nine months, it being the best locality for beetles I found during my twelve years' tropical collecting, and very good for other groups.

It was also in this place that I obtained numerous skins and skeletons of the orang-utan, as fully described in my "Malay Archipelago."

[1] They were sent by sailing ship round the Cape of Good Hope, the overland route being too costly for goods.

In another letter referring to the Dyaks, I say:— "The old men here relate with pride how many 'heads' they took in their youth; and though they all acknowledge the goodness of the present rajah, yet they think that if they were allowed to take a few heads, as of old, they would have better crops. The more I see of uncivilized people, the better I think of human nature on the whole, and the essential differences between civilized and savage man seem to disappear. Here we are, two Europeans, surrounded by a population of Chinese, Malays, and Dyaks. The Chinese are generally considered, and with some amount of truth, to be thieves, liars, and reckless of human life, and these Chinese are coolies of the lowest and least educated class, though they can all read and write. The Malays are invariably described as being barbarous and bloodthirsty; and the Dyaks have only recently ceased to think head-taking a necessity of their existence. We are two days' journey from Sarawak, where, though the government is nominally European, it only exists with the consent and by the support of the native population. Yet I can safely say that in any part of Europe where the same opportunities for crime and disturbance existed, things would not go so smoothly as they do here. We sleep with open doors, and go about constantly unarmed; one or two petty robberies and a little fighting have occurred among the Chinese, but the great majority of them are quiet, honest, decent sort of people."

In my next letter, a month later, I gave the following account of an interesting episode:—

"I must now tell you of the addition to my household of an orphan baby, a curious little half-nigger baby, which I have nursed now more than a month.

I will tell you presently how I came to get it, but must first relate my inventive skill as a nurse. The little innocent was not weaned, and I had nothing proper to feed it with, so was obliged to give it rice-water. I got a large-mouthed bottle, making two holes in the cork, through one of which I inserted a large quill so that the baby could suck. I fitted up a box for a cradle with a mat for it to lie upon, which I had washed and changed every day. I feed it four times a day, and wash it and brush its hair every day, which it likes very much, only crying when it is hungry or dirty. In about a week I gave it the rice-water a little thicker, and always sweetened it to make it nice. I am afraid you would call it an ugly baby, for it has a dark brown skin and red hair, a very large mouth, but very pretty little hands and feet. It has now cut its two lower front teeth, and the uppers are coming. At first it would not sleep alone at night, but cried very much; so I made it a pillow of an old stocking, which it likes to hug, and now sleeps very soundly. It has powerful lungs, and sometimes screams tremendously, so I hope it will live.

But I must now tell you how I came to take charge of it. Don't be alarmed; I was the cause of its mother's death. It happened as follows:—I was out shooting in the jungle and saw something up a tree which I thought was a large monkey or orangutan, so I fired at it, and down fell this little baby— in its mother's arms. What she did up in the tree of course I can't imagine, but as she ran about the branches quite easily, I presume she was a wild 'woman of the woods;' so I have preserved her skin and skeleton, and am trying to bring up her only daughter, and hope some day to introduce her to fashionable society at the Zoological Gardens. When its poor mother fell mortally wounded, the baby was plunged head over ears in a swamp about the consistence of pea-soup, and when I got it out looked very pitiful. It clung to me very hard when I carried it home, and having got its little hands unawares into my beard, it clutched so tight that I had great difficulty in extricating myself. Its mother, poor creature, had very long hair, and while she was running about the trees like a mad woman, the little baby had to hold fast to prevent itself from falling, which accounts for the remarkable strength of its little fingers and toes, which catch hold of anything with the firmness of a vice. About a week ago I bought a little monkey with a long tail, and as the baby was very lonely while we were out in the daytime, I put the little monkey into the cradle to keep it warm. Perhaps you will say that this was not proper. 'How could you do such a thing?' But, I assure you, the baby likes it exceedingly, and they are excellent friends. When the monkey wants to run away, as he often does, the baby clutches him by the tail or ears and drags him back; and if the monkey does succeed in escaping, screams violently till he is brought back again. Of course, baby cannot walk yet, but I let it crawl about on the floor to exercise its limbs; but it is the most wonderful baby I ever saw, and has such strength in its arms that it will catch hold of my trousers as I sit at work, and hang under my legs for a quarter of an hour at a time without being the least tired, all the time trying to suck, thinking, no doubt, it has got hold of its poor dear mother. When it finds no milk is to be had, there comes another scream, and I have to put it back in its cradle and give it 'Toby'—the little monkey—to hug, which quiets it immediately. From this short account you will see that my baby is no common baby, and I can safely say, what so many have said before with much less truth, 'There never was such a baby as my baby,' and I am sure nobody ever had such a dear little duck of a darling of a little brown hairy baby before."

In a letter dated Christmas Day, 1855, I gave my impressions of the Dyaks, and of Sir James Brooke, as follows:—

"I have now lived a month in a Dyak's house, and spent a day or two in several others, and I have been very much pleased with them. They are a very kind, simple, hospitable people, and I do not wonder at the great interest Sir James Brooke takes in them. They are more communicative and more cheerful than the American Indians, and it is therefore more agreeable to live with them. In moral character they are far superior to either the Malays or the Chinese, for though head-taking was long a custom among them, it was only as a trophy of war. In their own villages crimes are very rare. Ever since Sir James Brooke has been rajah, more than twelve years, there has only been one case of murder in a Dyak tribe, and that was committed by a stranger who had been adopted into the tribe. One wet day I produced a piece of string to show them how to play 'cat's cradle,' and was quite astonished to find that they knew it much better than I did, and could make all sorts of new figures I had never seen. They were also very clever at tricks with string on their fingers, which seemed to be a favourite amusement. Many of the remoter tribes think the rajah cannot be a man. They ask all sorts of curious questions about him— Whether he is not as old as the mountains; whether he cannot bring the dead to life; and I have no doubt, for many years after his death, he will be held to be a deity and expected to come back again.

I have now seen a good deal of Sir James, and the more I see of him the more I admire him. With the highest talents for government he combines in a high degree goodness of heart and gentleness of manner. At the same time, he has so much self-confidence and determination that he has put down with the greatest ease the conspiracies of one or two of the Malay chiefs against him. It is a unique case in the history of the world for a private English gentleman to rule over two conflicting races—a superior and an inferior—with their own consent, without any means of coercion, but depending solely upon them both for protection and support, while at the same time he introduces some of the best customs, of civilization, and checks all crimes and barbarous practices that before prevailed. Under his government 'running-a-muck,' so frequent in other Malay countries, has never taken place, and in a population of about 30,000 Malays, almost all of whom carry their *kris*, and were accustomed to revenge an insult with a stab, murders only occur once in several years. The people are never taxed except with their own consent, and in the manner most congenial to them, while almost the whole of the rajah's private fortune has been spent in the improvement of the country or for its benefit. Yet this is the man who has been accused in England of wholesale murder and butchery of unoffending tribes to secure his own power!"

In my next letter (from Singapore in February, 1856) I say—"I have now left Sarawak, where I began to feel quite at home, and may perhaps never return to it again, but I shall always look back with pleasure to my residence there and to my acquaintance with Sir James Brooke, who is a gentleman and a nobleman in the truest and best sense of those words."

While in Sarawak I wrote an article which formed my first contribution to the question of the origin of species. I sent it to *The Annals and Magazine of Natural History*, in which it appeared in the following September (1855). Its title was "On the Law which has regulated the Introduction of New Species," which law was briefly stated (at the end) as follows: "*Every species has come into existence coincident both in space and time with a pre-existing closely-allied species:*" This clearly pointed to some kind of evolution. It suggested the *when* and the *where* of its occurrence, and that it could only be through natural generation, as was also suggested in the "Vestiges "; but the *how* was still a secret only to be penetrated some years later.

Soon after this article appeared, Mr. Stevens, my agent,

wrote me that he had heard several naturalists express regret that I was "theorizing," when what we had to do was to collect more facts. After this, I had in a letter to Darwin expressed surprise that no notice appeared to have been taken of my paper, to which he replied that both Sir Charles Lyell and Mr. Edward Blyth, two very good men, specially called his attention to it. I was, however, rewarded later, when in Huxley's chapter, "On the Reception of the Origin of Species," contributed to the "Life and Letters," he referred to this paper as— "his powerful essay," adding—"On reading it afresh I have been astonished to recollect how small was the impression it made" (vol. ii. p. 185). The article is reprinted in my "Natural Selection and Tropical Nature."

In a letter to Bates, dated January 4, 1858, written on board the Dutch steamer which took me from Amboyna to Ternate, I wrote—"To persons who have not thought much on the subject I fear my paper on the 'Succession of Species' will not appear so clear as it does to you. That paper is, of course, merely the announcement of the theory, not its development. I have prepared the plan and written portions of a work embracing the whole subject, and have endeavoured to prove in detail what I have as yet only indicated. It was the promulgation of Forbes's theory of 'polarity' which led me to write and publish, for I was annoyed to see such an ideal absurdity put forth, when such a simple hypothesis will explain all the facts. I have been much gratified by a letter from Darwin, in which he says that he agrees with 'almost every word' of my paper. He is now preparing his great work on 'Species and Varieties,' for which he has been collecting materials twenty years. He may save me the trouble of writing more on my hypothesis, by proving that there is no difference in nature between the origin of species and of varieties; or he may give me trouble by arriving at another conclusion; but, at all events, his facts will be given for me to work upon. Your collections and my own will furnish most valuable material to illustrate and prove the universal applicability of the hypothesis. The connection between the succession of affinities and the geographical distribution of a group, worked out species by species, has never yet been shown as we shall be able to show it.

"In this archipelago there are two distinct faunas rigidly circumscribed, which differ as much as do those of Africa and South America, and more than those of Europe and North America, yet there is nothing on the map or on the face of the islands to mark their limits. The boundary line passes between islands closer together than others belonging to the same group. I believe the western part to be a separated portion of continental Asia, while the eastern is a fragmentary prolongation of a former west Pacific continent. In mammalia and birds the distinction is marked by genera, families, and even orders confined to one region; in insects by a number of genera, and little groups of peculiar species, the families of insects having generally a very wide or universal distribution."

This letter proves that at this time I had not the least idea of the nature of Darwin's proposed work, nor of the definite conclusions he had arrived at, nor had I myself any expectation of a complete solution of the great problem to which my paper was merely the prelude. Yet less than two months later that solution flashed upon me, and to a large extent marked out a different line of work from that which I had up to this time anticipated.

I finished the letter after my arrival at Ternate (January 25, 1858), and made the following observation: "If you go to the Andes I think you will be disappointed, at least in the *number of species*, especially of Coleoptera. My experience here is that the low grounds are *much* the most productive, though the mountains generally produce a few striking and brilliant species. "This rather hasty generalization is, I am inclined still to think, a correct one, at all events as regards the individual collector. I doubt if there is any mountain station in the world where so many species of butterflies can be collected within a walk as at Para, or more beetles than at my station in Borneo and Bates's at Ega. Yet it may be the case that many areas of about a hundred miles square in the Andes and in the Himalayas actually contain a larger number of species than any similar area in the lowlands of

the Amazon or of Borneo. In other parts of this letter I refer
to the work I hoped to do myself in describing, cataloguing,
and working out the distribution of my insects. I had in fact
been qitten by the passion for species and their description,
and if neither Darwin nor myself had hit upon "Natural
Selection," I might have spent the best years of my life
in this comparatively profitless work. But the new ideas
swept all this away. I have for the most part left others to
describe my discoveries, and have devoted myself to the
great generalizations which the laborious work of species-
describers had rendered possible. In this letter to Bates I
enclosed a memorandum of my estimate of the number of
distinct species of insects I had collected up to the time of
writing—three years and a half, nearly one year of which
had been lost in journeys, illnesses, and various delays. The
totals were as follows:—

Butterflies	620	species
Moths	2000	,,
Beetles	3700	,,
Bees, wasps, etc.	750	,,
Flies	660	,,
Bugs, cicadas, etc.	500	,,
Locusts, etc.	160	,,
Dragonflies, etc.	110	,,
Earwigs, etc	40	,,
Total	8540	species of Insects.

Having been unable to find a vessel direct to Macassar,
I took passage to Lombok, whence I was assured I should

easily reach my destination. By this delay, which seemed to me at the time a misfortune, I was enabled to make some very interesting collections in Bali and Lombok, two islands which I should otherwise never have seen. I was thus enabled to determine the exact boundary between two of the primary zoological regions, the Oriental and the Australian, and also to see the only existing remnant of the Hindu race and religion, and of the old civilization which had erected the wonderful ruined temples in Java centuries before the Mohammedan invasion of the archipelago.

After two months and a half in Lombok, I found a passage to Macassar, which I reached the beginning of September, and lived there nearly three months, when I left for the Aru Islands in a native prau. The country around Macassar greatly disappointed me, as it was perfectly flat and all cultivated as rice fields, the only sign of woods being the palm and fruit trees in the suburbs of Macassar and others marking the sites of native villages. I had letters to a Dutch merchant who spoke English as well as Malay and the Bugis language of Celebes, and who was quite friendly with the native rajah of the adjacent territory. Through his good offices I was enabled to stay at a native village about eight miles inland, where there were some patches of forest, and where I at once obtained some of the rare birds and insects peculiar to Celebes. After about a month I returned to Macassar, and found that I could obtain a passage to the celebrated Aru Islands, where at least two species of birds of paradise are found, and which had never been visited by an English collector. This was a piece of good fortune I had not expected, and it was especially so because the next six months would be wet in Celebes, while it would be the dry season in the Aru Islands. This journey was the most successful of any that I undertook, and as it is fully described in my book and no letters referring to it have been preserved, I shall say no more about it here.

The illustration opposite is from a photograph of a native house in the island of Wokan, which was given me by the late Professor Moseley of the *Challenger* expedition. It so closely resembles the hut in which I lived for a fortnight, and where I obtained my first King bird of paradise, that I feel sure it must be the same, especially as I saw no other like it.

NATIVE HOUSE IN ARU ISLANDS.

CHAPTER XIII

THE MALAY ARCHIPELAGO

(1858–1862)

AFTER returning to Macassar, where I spent three months collecting, I visited Amboyna, staying there a month, and arrived at Ternate, January 8, 1858.

It was while waiting at Ternate in order to get ready for my next journey, and to decide where I should go, that the idea already referred to occurred to me. It has been shown how, for the preceding eight or nine years, the great problem of the origin of species had been continually pondered over, and how my varied observations and study had been made use of to lay down the foundation for its full discussion and elucidation. My paper written at Sarawak rendered it certain to my mind that the change had taken place by natural succession and descent—one species becoming changed either slowly or rapidly into another. But the exact process of the change and the causes which led to it were absolutely unknown and appeared almost inconceivable.

The great difficulty was to understand how, if one species was gradually changed into another, there continued to be so many quite distinct species, so many which differed from their nearest allies by slight yet perfectly definite and constant characters.

At the time in question I was suffering from a sharp attack of intermittent fever, and every day during the cold and succeeding hot fits had to lie down for several hours, during which time I had nothing to do but to think over any subjects then particularly interesting me. One day something brought to my recollection Malthus's "Principles of Population," which I had read about twelve years before. I thought of his clear exposition of "the positive checks to increase"—disease, accidents, war, and famine—which keep down the population of savage races to so much lower an average than that of more civilized peoples. It then occurred to me that these causes or their equivalents are continually acting in the case of animals also; and as animals usually breed much more rapidly than does mankind, the destruction every year from these causes must be enormous in order to keep down the numbers of each species, since they evidently do not increase regularly from year to year, as otherwise the world would long ago have been densely crowded with those that breed most quickly. Vaguely thinking over the enormous and constant destruction which this implied, it occurred to me to ask the question, Why do some die and some live? And the answer was clearly, that on the whole the best fitted live. From the effects of disease the most healthy escaped; from enemies, the strongest, the swiftest, or the most cunning; from famine, the best hunters or those with the best digestion; and so on. Then it suddenly flashed upon me that this self-acting process would necessarily *improve the race*, because in every generation the inferior would inevitably be killed off and the superior would remain—that is, *the fittest would survive*. Then at once I seemed to see the whole effect of this, that when changes of land and sea, or of climate, or of food-supply, or of enemies occurred—and we know that such changes have always been taking place—in conjunction with the amount of individual variation that my experience as a collector had shown me to exist, then all the changes necessary for the adaptation of the species to the changing conditions would be brought about; and as great changes in the environment are always slow, there would be ample time for the change to be effected by the survival of the best fitted in every generation. In this way each part of an animal's organization could be modified exactly as required, and in the very process of this modification the unmodified would die out, and thus the *definite* characters and the clear *isolation* of each new species would be explained. The more I thought over it the more I became convinced that I had at length found the long-sought-for law of nature that solved the problem of the origin of species. For the next hour I thought over the deficiencies in the theories of Lamarck and of the author of the "Vestiges," and I saw that my new theory supplemented these views and obviated every important difficulty. I waited anxiously for the termination of my fit so that I might at once make notes for a paper on the subject. The same evening I did this pretty fully, and on the two succeeding evenings wrote it out carefully in order to send it to Darwin by the next post, which would leave in a day or two.

I wrote a letter to him in which I said that I hoped the idea would be as new to him as it was to me, and that it would supply the missing factor to explain the origin of species. I asked him, if he thought it sufficiently important, to show it to Sir Charles Lyell, who had thought so highly of my former paper.

The effect of my paper upon Darwin was at first almost paralyzing. He had, as I afterwards learnt, hit upon the same idea as my own twenty years earlier, and had occupied himself during the whole of that long period in study and experiment, and in sketching out and partly writing a great work, to show how the new principle would serve to explain almost all the chief phenomenon and characters of living things in their relation to each other.

So early as 1844 he had shown portions of this work to Sir Charles Lyell and Dr. Joseph Hooker, who had been greatly struck by it, and who were thenceforth his only confidants in the secret of his new idea, which from the analogy of the breeder's selection of the most suitable animals or plants in order to produce new varieties, he termed "natural selection." Sir Charles Lyell had frequently urged him to publish an outline of his views, saying—"If you don't, some one else will hit upon it, and you will be forestalled." On receiving my paper he wrote to Sir Charles—"Your words have come true with a vengeance—that I should be forestalled. I never saw a more striking coincidence.... So all my originality, whatever it may amount to, will be smashed, though my book, if it will ever have any value, will not be deteriorated, as all the labour consists in the application of the theory "("Life and Letters of Charles Darwin," ii. p. 116).

Darwin was naturally much troubled, and did not know how to act, declaring in a later letter to Sir Charles Lyell—"I would far rather burn my whole book, than that he (Wallace) or any other man should think that I had behaved in a paltry spirit." He therefore left the matter in the hands of his two friends, and they determined (on their own responsibility) that my essay, together with extracts from Darwin's MSS., which they had seen many years before, should be read before the Linnean Society and published in its "Journal." The joint papers were read on July 1, 1858, Dr. Hooker and Sir Charles Lyell being present. After Darwin's death the former wrote a short account of this event for the "Life and Letters," in which he said: "The interest excited was intense, but the subject was too novel and too ominous for the old school to enter the lists before armouring. After the meeting it was talked over with bated breath: Lyell's approval, and perhaps in a small way mine, as his lieutenant in the affair, rather overawed the Fellows, who would otherwise have flown out against the doctrine" ("Life and Letters," ii. p. 126).

Both Darwin and Dr. Hooker wrote to me in the most kind and courteous manner, informing me of what had been done, of which they hoped I would approve. Of course I not only approved, but felt that they had given me more honour and credit than I deserved, by putting my sudden intuition— hastily written and immediately sent off for the opinion of Darwin and Lyell—on the same level with the prolonged labours of Darwin, who had reached the same point twenty years before me, and had worked continuously during that long period in order that he might be able to present the theory to the world with such a body of systematized facts and arguments as would almost compel conviction.

In a later letter, Darwin wrote that he owed much to me and his two friends, adding: "I almost think that Lyell would have proved right, and that I should never have completed my larger work." I think, therefore, that I may have the satisfaction of knowing that by writing my article and sending it to Darwin, I was the unconscious means of leading him to concentrate himself on the task of drawing up what he termed "an abstract" of the great work he had in preparation, but which was really a large and carefully written volume—the celebrated "Origin of Species," published in November, 1859. The remainder of the story of my relations with Darwin will be found in later chapters of this book. I will now continue the account of my travels.

During the first months of my residence at Ternate I made two visits to different parts of the large island of Gilolo, where my hunters obtained a number of very fine birds, but owing to the absence of good virgin forest and my own ill-health, I obtained very few insects. At length, on March 25, I obtained a passage to Dorey Harbour, on the north coast of New Guinea, in a trading schooner, which left me there, and called for me three or four months later to bring me back to Ternate. I was the first European who had lived alone on this great island; but partly owing to an accident which confined me to the house for a month, and partly because the locality was not a good one, I did not get the rare species of birds of paradise I had expected. I obtained, however, a number of new and rare birds and a fine collection of insects, though not so many of the larger and finer kinds as I expected. The weather had been unusually wet, and the place was unhealthy. I had four Malay servants with me, three of whom had fever as well as myself, and one of my hunters died, and though I should have liked to stay longer, we were all weak or unwell, and were very glad when the schooner arrived and took us back to Ternate. Here wholesome food and a comfortable house soon restored us to good health.

I now received letters informing me of the reception of the paper on "The Tendency of Varieties to depart indefinitely from the Original Type," which I had sent to Darwin, and in a letter home I thus refer to it: "I have received letters from Mr. Darwin and Dr. Hooker, two of the most eminent naturalists in England, which have highly gratified me. I sent Mr. Darwin an essay on a subject upon which he is now writing a great work. He showed it to Dr. Hooker and Sir Charles Lyell, who thought so highly of it that they had it read before the Linnean Society. This insures me the acquaintance of these eminent men on my return home."

The next two years were spent principally at Ternate, Batchian, Menado in Celebes, and Amboyna.

In February, 1860, I left the last-named place with the intention of again visiting the Ké Islands. After immense difficulties I reached Goram, about fifty miles beyond the east end of Ceram, where I purchased a boat and started for Ké; but after getting half-way, the weather was so bad and the winds so adverse that I was obliged to return to the Matabello Islands, and thence by way of Goram and the north coast of Ceram to the great island of Waigiou. This was a long and most unfortunate voyage. I found there, however, what I chiefly went for—the rare red bird of paradise (*Paradisea rubra*); but during the three months I lived there, often with very little food, I obtained only about seventy species of birds, mostly the same as those from New Guinea, though a few species of parrots, pigeons, kingfishers, and other birds were new. Insects were never abundant, but by continued perseverance I obtained rather more species of both butterflies and beetles than at New Guinea, though fewer, I think, of the more showy kinds.

The voyage from Waigiou back to Ternate was again most tedious and unfortunate, occupying thirty-eight days, whereas with reasonably favourable weather it should not have required more than ten or twelve. Taking my whole voyage in this canoe from Goram to Waigiou and Ternate, I thus summarize my account of it in my "Malay Archipelago": "My first crew ran away in a body; two men were lost on a desert island, and only recovered a month later after twice sending in search of them; we were ten times run aground on coral reefs; we lost four anchors; our sails were devoured by rats; our small boat was lost astern; we were thirty-eight days on a voyage which should not have taken twelve; we were many times short of food and water; we had no compass-lamp owing to there being not a drop of oil in Waigiou when we left; and, to crown all, during our whole voyage from Goram by Ceram to Waigiou, and from Waigiou to Ternate, occupying in all seventy-eight days (or only twelve days short of three months), all in what was supposed to be the favourable season, we had *not one single day of fair wind*. We were always close braced up, always struggling against wind, currents, and leeway, and in a vessel that would scarcely sail nearer than eight points from the wind! Every seaman will admit that my first (and last) voyage in a boat of my own was a very unfortunate one."

On again returning to Ternate I wrote the following letter to Bates, giving my opinion of Darwin's "Origin of Species," and referring to the subject of the geographical distribution of animals in which I was much interested:—

"Ternate, December 24, 1860.

DEAR BATES,

Many thanks for your long and interesting letter. I have myself suffered much in the same way as you describe, and I think more severely. The kind of *tædium vitæ* you mention I also occasionally experience here. I impute it to a too monotonous existence.

I know not how, or to whom, to express fully my admiration of Darwin's book. To *him* it would seem flattery, to others self-praise; but I do honestly believe that with however much patience I had worked and experimented on the subject, I could *never have approached* the completeness of his book, its vast accumulation of evidence, its overwhelming argument, and its admirable tone and spirit. I really feel thankful that it has *not* been left to me to give the theory to the world. Mr. Darwin has created a new science and a new philosophy; and I believe that never has such a complete illustration of a new branch of human knowledge been due to the labours and researches of a single man. Never have such vast masses of widely scattered and hitherto quite unconnected facts been combined into a system and brought to bear upon the establishment of such a grand and new and simple philosophy.

I am surprised at your joining the north and south banks of the lower Amazon into one region. Did you not find a sufficiency of distinct species at Obydos and Barra to separate them from Villa Nova and Santarem? I am now convinced that insects, on the whole, do not give such true indications of zoological geography as birds and mammals, because, first, they have such immensely greater means of dispersal across rivers and seas; second, because they are so much more influenced by surrounding circumstances; and third, because the species seem to change more quickly, and therefore disguise a comparatively recent identity. Thus the insects of adjacent regions, though originally distinct, may become rapidly amalgamated, or portions of the same region may come to be inhabited by very distinct insect-faunas owing to differences of soil, climate, etc. This is strikingly shown here, where the insect-fauna from Malacca to New Guinea has a very large amount of characteristic uniformity, while Australia, from its distinct climate and vegetation, shows a wide difference. I am inclined to think, therefore, that a preliminary study of, first, the mammals, and then the birds, is indispensable to a correct understanding of the geographical and physical changes on which the present insect-distribution depends.hellip

In a day or two I leave for Timor, where, if I am lucky in finding a good locality, I expect some fine and interesting insects."

My last two letters before coming home were written in the wilds of Sumatra to Bates and Silk.

An extract from the letter to Bates will give some idea of the difficulty I had in finding good collecting places.

"I am here making what I intend to be my last collections, but am doing very little in insects, as it is the wet season and all seems dead. I find in those districts where the seasons are strongly contrasted the good collecting time is very limited—only about a month or two at the beginning of the dry, and a few weeks at the commencement of the rains. It is now two years since I have been able to get any beetles, owing to bad localities and bad weather, so I am becoming disgusted. When I do find a good place it is generally very good, but such are dreadfully scarce. In Java I had to go forty miles in the eastern part and sixty miles in the western to reach a bit of forest, and then I got scarcely anything. Here I had to come a hundred miles inland, by Palembang, and though in the very centre of Eastern Sumatra, the forest is only in patches, and it is the height of the rains, so I get nothing. A longicorn is a rarity, and I suppose I shall not have as many species in two months as I have obtained in three or four days in a really good locality. I am getting, however, some sweet little blue butterflies (*Lycœnidœ*), which is the only thing that keeps up my spirits."

An extract from my letter to Silk will be of more interest to most of my readers: "I am here in one of the places unknown to the Royal Geographical Society, situated in the very centre of East Sumatra, about one hundred miles from the sea in three directions. It is the height of the wet season, and the rain pours down strong and steady, generally all night and half the day. Bad times for me, but I walk out regularly three or four hours every day, picking up what I can, and generally getting some little new or rare or beautiful thing to reward me. This is the land of the two-horned rhinoceros, the elephant, the tiger, and the tapir; but they all make themselves very scarce, and beyond their tracks and their dung, and once hearing a rhinoceros *bark* not far off, I am not aware of their existence. This, too, is the very land of monkeys; they swarm about the villages and plantations, long-tailed and short-tailed, and with no tail at all, white, black, and grey; they are eternally racing about the tree-tops, and gambolling in the most amusing manner. The way they jump is amazing. They throw themselves recklessly through the air, apparently sure, with one or other of their four hands, to catch hold of something. I estimated one jump by a long-tailed white monkey at thirty feet horizontal, and sixty feet vertical, from a high tree on to a lower one; he fell through, however, so great was his impetus, on to a lower branch, and then, without a moment's stop, scampered away from tree to tree, evidently quite pleased with his own pluck. When I startle a band, and one leader takes a leap like this, it is amusing to watch the others—some afraid and hesitating on the brink till at last they pluck up courage, take a run at it, and often roll over in the air with their desperate efforts. Then there are the long-armed apes, who never walk or run upon the trees, but travel altogether by their long arms, swinging themselves from bough to bough in the easiest and most graceful manner possible.

"But I must leave the monkeys and turn to the men, who will interest you more, though there is nothing very remarkable in them. They are Malays, speaking a curious, half-unintelligible Malay dialect— Mohammedans, but retaining many pagan customs and superstitions. They are very ignorant, very lazy, and live almost absolutely on rice alone, thriving upon it, however, just as the Irish do, or did, upon potatoes. They were a bad lot a few years ago, but the Dutch have brought them into order by their admirable system of supervision and government. By-the-by, I hope you have read Mr. Money's book on Java. It is well worth while, and you will see that I had come to the same conclusions as to Dutch colonial government from what I saw in Menado. Nothing is worse and more absurd than the sneering prejudiced tone in which almost all English writers speak of the Dutch government in the East. It never has been worse than ours has been, and it is now very much better; and what is greatly to their credit and not generally known, they take nearly the same pains to establish order and good government in those islands and possessions which are an annual loss to them, as in those which yield them a revenue. I am convinced that their system is *right* in principle, and ours *wrong,* though, of course, in the practical working there may and must be defects; and among the Dutch themselves, both in Europe and the Indies, there is a strong party against the present system, but that party consists mostly of merchants and planters, who want to get the trade and commerce of the country made free, which in my opinion would be an act of suicidal madness, and would, moreover, seriously injure instead of benefiting the natives.

Personally, I do not much like the Dutch out here, or the Dutch officials; but I cannot help bearing witness to the excellence of their government of native races, gentle yet firm, treating their manners, customs, and prejudices with respect, yet introducing everywhere European law, order, and industry."

"Singapore, January 20, 1862.

I cannot write more now. I do not know how long I shall be here; perhaps a month. Then, ho! for England!"

While waiting at Singapore for the steamer to take me

home I purchased two living specimens of the smaller bird of paradise. They were in a large cage, and the price asked was enormous. As they had never been seen alive in Europe I at once secured them, and had a great deal of trouble with them on my journey home. I had first to make an arrangement for a place to stand the large cage on deck. A stock of food was required, which consisted chiefly of bananas; but to my surprise I found that they would eat cockroaches greedily, and as these abound on every ship in the tropics, I hoped to be able to obtain a good supply. Every evening I went to the storeroom in the fore part of the ship, where I was allowed to brush the cockroaches into a biscuit tin.

The journey to Suez offered no particular incident, and the birds continued in good health; as did two or three lories I had brought. But with the railway journey to Alexandria difficulties began. It was in February, and the night was clear and almost frosty. The railway officials made difficulties, and it was only by representing the rarity and value of the birds that I could have the cage placed in a box-truck. When we got into the Mediterranean the weather became suddenly cold, and worse still, I found that the ship was free from cockroaches. As I thought that animal food was perhaps necessary to counteract the cold, I felt afraid for the safety of my charge, and determined to stay a fortnight at Malta in order to reach England a little later, and also to lay in a store of the necessary food. I accordingly arranged to break my voyage there, went to a hotel, and found that I could get unlimited cockroaches at a baker's close by.

At Marseilles I again had trouble, but at last succeeded in getting them placed in a guard's van, with permission to enter and feed them *en route*. Passing through France it was a sharp frost, but they did not seem to suffer; and when we reached London I was glad to transfer them into the care of Mr. Bartlett, who conveyed them to the Zoological Gardens.

Thus ended my Malayan travels.

CHAPTER XIV

LIFE IN LONDON, 1862–1871—
SCIENTIFIC AND LITERARY WORK

ON reaching London in the spring of 1862 I went to live with my brother-in-law, Mr. Thomas Sims, who had a photographic business in Westbourne Grove. Here, in a large empty room at the top of the house, I brought together all the collections which I had reserved for myself and which my agent, Mr. Stevens, had taken care of for me. I found myself surrounded by a quantity of packing-cases and store-boxes, the contents of many of which I had not seen for five or six years, and to the examination and study of which I looked forward with intense interest.

From my first arrival in the East I had determined to keep a complete set of certain groups from every island or distinct locality which I visited for my own study on my return home, as I felt sure they would afford me very valuable materials for working out the geographical distribution of animals in the archipelago, and also throw light on various other problems. These various sets of specimens were sent home regularly with the duplicates for sale, but either packed separately or so distinctly marked "Private" that they could be easily put aside till my return home. The groups thus reserved were the birds, butterflies, beetles, and land-shells, and they amounted roughly to about three thousand bird skins of about a thousand species, and, perhaps, twenty thousand beetles and butterflies of about seven thousand species.

For the next month I was fully occupied in the unpacking and arranging of my collections, while I usually attended the evening meetings of the Zoological, Entomological, and Linnean Societies, where I met many old friends and made several new ones, and greatly enjoyed the society of people interested in the subjects that now had almost become the business of my life.

As soon as I began to study my birds I had to pay frequent visits to the bird-room of the British Museum, then in charge of Mr. George Robert Gray, who had described many of my discoveries as I sent them home, and also to the library of the Zoological Society to consult the works of the older ornithologists. In this way the time passed rapidly, and I became so interested in my various occupations, and saw so many opportunities for useful and instructive papers on various groups of my birds and insects, that I came to the conclusion to devote myself for some years to this work, and to put off the writing of a book on my travels till I could embody in it all the more generally interesting results derived from the detailed study of certain portions of my collections. This delay turned out very well, as I was thereby enabled to make my book not merely the journal of a traveller, but also a fairly complete sketch of the whole of the great Malayan Archipelago from the point of view of the philosophic naturalist. The result has been that it long continued to be the most popular of my books, and that even now, forty years after its publication, its sale is equal to that of any of the others.

During the succeeding five years I continued the study of my collections, writing many papers, of which more than a dozen related to birds, some being of considerable length and involving months of continuous study. But I also wrote several on physical and zoological geography, six on various questions of anthropology, and five or six on special applications of the theory of natural selection. I also began working at my insect collections, on which I wrote four rather elaborate papers. As several of these papers discussed matters of considerable interest and novelty, I will here give a brief summary of the more important of them in the order in which they were written.

The first of these, read in January, 1863, at a meeting of the Zoological Society, was on my birds from Bouru, and was chiefly important as showing that this island was undoubtedly one of the Moluccan group, every bird found there which was not widely distributed being either identical with or closely allied to Moluccan species, while none had special affinities with Celebes. It was clear, then, that this island formed the most westerly outlier of the Moluccan group.

My next paper of importance, read before the same society in the following November, was on the birds of the chain of islands extending from Lombok to the great island of Timor. I gave a list of one hundred and eighty-six species of birds, of which twenty-nine were altogether new; but the special importance of the paper was that it enabled me to mark out precisely the boundary line between the Indian and Australian zoological regions, and to trace the derivation of the rather peculiar fauna of these islands, partly from Australia and partly from the Moluccas, but with a strong recent migration of Javanese species due to the very narrow straits separating most of the islands from each other. The following table will serve to illustrate this:—

	Lombok.	Flores.	Timor.
Species derived from Java	34	28	17
Species derived from Australia	7	14	36

This table shows how two streams of immigration have entered these islands, the one from Java diminishing in intensity as it passed on farther and farther to Timor; the other from Australia entering Timor and diminishing still more rapidly towards Lombok. This indicates, as its geological structure shows, that Timor is the older island and that it received immigrants from Australia at a period when, probably, Lombok and Flores had not come into existence or were uninhabitable.

Two other papers dealt with the parrots and the pigeons of the whole archipelago, and are among the most important of my studies of geographical distribution. That on parrots was written in 1864, and read at a meeting of the Zoological Society in June.

The peculiarities of distribution and coloration in two such very diverse groups of birds interested me greatly, and I endeavoured to explain them in accordance with the laws of natural selection. In the paper on Pigeons (published in *The Ibis* of October, 1865) I suggest that the excessive development of both these groups in the Moluccas and the Papuan islands has been due primarily to the total absence of arboreal, carnivorous, or egg-destroying mammals, especially of the whole monkey tribe, which in all other tropical forest regions are exceedingly abundant, and are very destructive to eggs and young birds. I also point out that there are here comparatively few other groups of fruit-eating birds like the extensive families of chatterers, tanagers, and toucans of America, or the barbets, bulbuls, finches, starlings, and many other groups of India and Africa, while in all those countries monkeys, squirrels, and other arboreal mammals consume enormous quantities of fruits. It is clear, therefore, that in the Australian region, especially in the forest-clad portions of it, both parrots and pigeons have fewer enemies and fewer competitors for food than in other tropical regions, the result being that they have had freer scope for development in various directions leading to the production of forms and styles of colouring unknown elsewhere. It is also very suggestive that the only other country in which *black* pigeons and *black* parrots are found is Madagascar, an island where also there are neither monkeys nor squirrels, and where arboreal carnivora or fruit-eating birds are very scarce. The satisfactory solution of these curious facts of distribution gave me very great pleasure, and I am not aware that the conclusions I arrived at have been seriously objected to.

Before I had written these two papers I had begun the study of my collection of butterflies, and in March, 1864, I read before the Linnean Society a rather elaborate paper on "The Malayan Papilionidæ, as illustrating the Theory of Natural Selection."

I may state for the information of non-entomological readers that the Papilionidæ form one of the most extensive families of butterflies, and from their large size, elegant forms, and splendid colours were considered by all the older writers to be the princes of the whole lepidopterous order. In coloration they are wonderfully varied. The ground colour is very frequently black, on which appear bands, spots, or large patches of brilliant colours—pale or golden yellow, rich crimsons or gorgeous metallic blues and greens, which colours sometimes spread over nearly the whole wing surface. Some are thickly speckled with golden green dots and adorned with large patches of intense metallic green or azure blue, others are simply black and white in a great variety of patterns, many very striking and beautiful, while others again have crimson or golden patches, which when viewed at certain angles change to quite different opalescent hues, unsurpassed by the rarest gems.

But it is not this grand development of size and colour that constitutes the attraction of these insects to the student of evolution, but the fact that they exhibit, in a remarkable degree, almost every kind of variation, as well as some of the most beautiful examples of polymorphism and of mimicry. Besides these features, the family presents us with examples of differences of size, form, and colour, characteristic of certain localities, which are among the most singular and mysterious phenomena known to naturalists. A short statement of the nature of these phenomena will be useful to show the great interest of the subject.

In all parts of the world there are certain insects which, from a disagreeable smell or taste, are rarely attacked or devoured by enemies. Such groups are said to be "protected," and they almost always have distinctive and conspicuous colours. In the Malay Archipelago there are several groups of butterflies which have this kind of protection; and one group —the Euplæas—is coloured black, with rich blue glosses and ornamented with white bands or spots. These are excessively abundant, and having few enemies they fly slowly. Now there are also several different kinds of papilios, which in colour are so exactly like these, that when on the wing they cannot be distinguished, although they frequent the same places and are often found intermingled. Other protected butterflies are of paler colours with dark stripes, and these are also closely imitated by other papilios. Altogether there are about fifteen species which thus closely resemble protected butterflies externally, although in structure and transformations they have no affinity with them. In some cases both sexes possess this resemblance, or "mimicry," as it is termed, but most frequently it is the female only that is thus modified, especially when she lays her eggs on low-growing plants; while the male, whose flight is stronger and who can take care of himself, does not possess it, and is often so different from his mate as to have been considered a distinct species.

This leads us to the phenomenon of dimorphism and polymorphism, in which the females of one species present two or three different forms. Several such cases occur in the Malay Archipelago, in which there are two distinct kinds of females, sometimes even three, to a single male, which differs from either of them. In one case four females are known to one male, though only two of them appear to occur in one locality. These have been almost always described as distinct species, but observation has now proved them to be one, and it has further been noticed that each of the females, which are very unlike the male, resembles more or less closely some "protected" species. It has also been proved by experimental breeding that eggs laid by any one of these females are capable of producing butterflies of all the different forms, which in the few cases recorded are quite distinct from each other, without intermediate gradations.

This brief outline of the paper will, perhaps, enable my readers to understand the intense interest I felt in working out all these strange phenomena, and showing how they could almost all be explained by that law of "Natural Selection" which Darwin had discovered many years before, and which I had also been so fortunate as to hit upon.

The series of papers on birds and insects now described, together with others on the physical geography of the archipelago and its various races of man, furnished me with the necessary materials for that general sketch of the natural history of the islands and of the various interesting problems which arise from its study, which has made my "Malay Archipelago" the most popular of my books. At the same time it opened up so many fields of research as to render me indisposed for further technical work in the mere description of my collections, which I should certainly never have been able to complete. I therefore now began to dispose of various portions of my insects to students of special groups, who undertook to publish lists of them with descriptions of the new species, reserving for myself only a few boxes of duplicates to serve as mementoes of the exquisite or fantastic organisms which I had procured during my eight years' wanderings.

Having thus prepared the way by these preliminary studies, I devoted the larger portion of my time in the years 1867 and 1868 to writing my "Malay Archipelago" I had previously read what works I could procure on the islands, and had made numerous extracts from the old voyagers on the parts I myself was acquainted with. These added much to the interest of my own accounts of the manners and character of the people, and by means of a tolerably full journal and the various papers I had written, I had no difficulty in going steadily on with my work. As my publishers wished the book to be well illustrated, I had to spend a good deal of time in deciding on the plates and getting them drawn, either from my own sketches, from photographs, or from actual specimens, and having obtained the services of the best artists and wood engravers then in London, the result was, on the whole, satisfactory.

The book was published in 1869, but during its progress, and while it was slowly passing through the press, I wrote several important papers, among which was one in the *Quarterly Review* for April, 1869, on "Geological Climates and the Origin of Species," which was in large part a review and eulogy of Sir Charles Lyell's great work, "The Principles of Geology," which greatly pleased him as well as Darwin. A considerable part of this article was devoted to a discussion of Mr. Croll's explanation of the glacial epoch, and, by a combination of his views with those of Lyell on the great effect of changed distribution of sea and land, or of differences in altitude, I showed how we might arrive at a better explanation than either view by itself could give us. As the article was too long, a good deal of it had to be cut out, but it served as the foundation for my more detailed examination of the whole question when writing my "Island Life," twelve years later.

As soon as the proofs of the "Malay Archipelago" were out of my hands, I began the preparation of a small volume of my scattered articles dealing with various aspects of the theory of Natural Selection.

Many of these had appeared in little-known periodicals, and were now carefully revised, or partially rewritten, while two new ones were added. The longest article, occupying nearly a quarter of the volume, was one which I had written in 1865–6, but which was not published (in the *Westminster Review*) till July, 1867, and was entitled "Mimicry, and other Protective Resemblances among Animals." In this article I endeavoured to give a general account of the whole subject of protective resemblance, of which theory, what was termed by Bates "mimicry," is a very curious special case. I called attention to the wide extent of the phenomenon, and showed that it pervades animal life from mammals to fishes and through every grade of the insect tribes. I pointed out that the whole series of phenomena depend upon the great principle of the utility of every character, upon the need of protection or of concealment by almost all animals, and upon the known fact that no character is so variable as colour, and that therefore concealment has been most easily obtained by colour modification.

Two other articles which may be just mentioned are those entitled "A Theory of Birds' Nests" and "The Limits of Natural Selection applied to Man." In the first I pointed out the important relation that exists between concealed nests and the bright colours of female birds, leading to conclusions adverse to Mr. Darwin's theory of colours and ornaments in the males being the result of female choice. In the other (the last in the volume) I apply Darwin's principle of natural selection, acting solely by means of "utilities," to show that certain physical modifications and mental faculties in man could not have been acquired through the preservation of useful variations, because there is some direct evidence to show that they *were not* and *are not* useful in the ordinary sense, or, as Professor Lloyd Morgan well puts it, not of "life-preserving value," while there is absolutely no evidence to show that they were so. In reply, Darwin appealed to the effects of female choice in developing these characteristics, of which, however, not a particle of evidence is to be found among existing savage races.

As it was during the ten years of which I have now sketched my scientific and literary work that I saw most of my various scientific friends and acquaintances, and it was also in this period that the course of my future life and work was mainly determined, I will give a short summary of my more personal affairs, together with a few recollections of those friends with whom I became most familiar.

In the spring of 1865 I took a small house for myself and my mother, in St. Mark's Crescent, Regent's Park, quite near the Zoological Gardens, and within a pleasant walk across the park of the society's library in Hanover Square, where I had to go very often to consult books of reference. Here I lived five years, having Dr. W. B. Carpenter for a near neighbour, and it was while living in this house that I saw most of my few scientific friends.

About this time my dear friend, Dr. Richard Spruce, came home from Peru in very weak health, and, after staying a short time in London, went to live at Hurstpierpoint, in Sussex, in order to be near Mr. William Mitten, then the greatest English authority on mosses, who had undertaken to describe his great collections from South America.

During the summer and autumn I often went to Hurstpierpoint to enjoy the society of my friend, and thus became intimate with Mr. Mitten and his family. Mr. Mitten was an enthusiastic botanist and gardener, and knew every wild plant in the very rich district which surrounds the village, and all his family were lovers of wild flowers. I remember my delight, on the occasion of my first or second visit there, at seeing a vase full of the delicate and fantastic flowers of the large butterfly-orchis and the curious fly-orchis, neither of which I had ever seen before, and which I was surprised to hear were abundant in the woods at the foot of the downs. It was an immense delight to me to be taken to these woods, and to some fields on the downs where the bee-orchis and half a dozen other species grew abundantly, with giant cowslips nearly two feet high, the dyers' broom, and many other interesting plants. The richness of this district may be judged by the fact that within a walk more than twenty species of orchises have been found. This similarity of taste led to a close intimacy, and in the spring of the following year I was married to Mr. Mitten's eldest daughter, then about eighteen years old.

After a week at Windsor we came to live in London, and in early autumn went for a month to North Wales, staying at Llanberris and Dolgelly. I took with me Sir Andrew Ramsay's little book on "The Old Glaciers of Switzerland and North Wales," and thoroughly enjoyed the fine examples of ice-groovings and striations, smoothed rock-surfaces, *roches moutonnées,* moraines, perched blocks, and rock-basins, with which the valleys around Snowdon abound. Every day revealed some fresh object of interest as we climbed among the higher *cwms* of Snowdon; and from what I saw during that first visit the Ice Age became almost as much a reality to me as any fact of direct observation. Every future tour to Scotland, to the lake district, or to Switzerland became doubled in interest. I read a good deal of the literature of the subject, and have, I believe, in my later writings been able to set forth the evidence in favour of the glacial origin of lake-basins more forcibly than it has ever been done before.

In 1867 I spent the month of June in Switzerland with my wife, staying at Champery, opposite the beautiful Dent du Midi, where at first we were the only visitors in a huge new hotel, but for the second week had the company of an English clergyman, his wife, and son. We greatly enjoyed the beautiful subalpine flowers then in perfection.

We then went by Martigny over the St. Bernard, reaching the hospice after dark through deep snow, and next day walked down to Aosta, a place which had been recommended to me by Mr. William Mathews, a well-known Alpine climber. It was a very hot place, and its chief interest to us was an excursion on mules to the Becca de Nona, which took us a long day, going, up by the easiest and descending the most precipitous road—the latter a mere staircase of rock. The last thousand feet of the mountain I walked up alone, and was highly delighted with the summit and the wonderful scene of fractured rocks, ridges, and peaks all around, but more especially with the summit itself, hardly so large as that of Snowdon and exhibiting far grander precipices and rockmasses, all in a state of visible degradation, and showing how powerfully the atmospheric forces of denudation are in constant action at this altitude — 10,380 feet. Hardly less interesting were the charming little alpine plants in the patches of turf and the crevices in the rocks, among which were two species of the exquisite Androsaces, the true gems of the primrose tribe. I also one day took a lonely walk up a wild valley which terminated in the glacier that descends from Mount Emilius; and on another day we drove up the main valley to Villeneuve, and then walked up a little way into the Val Savaranches. This is one of those large open valleys which have been the outlet of a great glacier, and in which the subglacial torrent has cut a deep narrow chasm through hard rocks at its termination, through which the river now empties itself into the main stream of the Dora Baltea. This was the first of the kind I had specially noticed, though I had seen the Gorge of the Trient on my first visit to Switzerland at a time when I had barely heard of the glacial epoch.

Returning over the St. Bernard we went to Interlaken and Grindelwald, saw the glaciers there, and then went

A. R. WALLACE. AGED 46.
(*Publication of Malay Archipelago.*)

over the Wengern Alp, staying two days at the hotel to see the avalanches and botanize among the pastures and moraines. Then down to Lauterbrunnen to see the Staubbach, and thence home.

As I had found that amid the distractions and excitement of London, its scientific meetings, dinner parties and sight-seeing, I could not settle down to work at the more scientific chapters of my "Malay Archipelago," I let my house in London for a year, from Midsummer, 1867, and went to live with my wife's family at Hurstpierpoint. There, in perfect quiet, and with beautiful fields and downs around me, I was able to work steadily, having all my materials already prepared. Returning to London in the summer of 1868, I was fully occupied in arranging for the illustrations and correcting the proofs. The work appeared early in the new year, and my volume on "Natural Selection" in the following March.

I may here state that although the proceeds of my eight years' collecting in the East brought me in a sufficient income to live quietly as a single man, I was always on the look-out for some permanent congenial employment which would yet leave time for the study of my collections. The possibility of ever earning anything substantial either by lecturing or by writing never occurred to me. My deficient organ of language prevented me from ever becoming a good lecturer or having any taste for it, while the experience of my first work on "The Amazon" did not encourage me to think that I could write anything that would much more than pay expenses. The first vacancy that occurred was the assistant secretaryship of the Royal Geographical Society, for which Bates and myself were candidates. Bates had just published his "Naturalist on the Amazon," and was, besides, much better qualified than myself by his business experience and his knowledge of German, which he had taught himself when abroad. Besides, the confinement and the London life would, I am sure, have soon become uncongenial to me, and would, I feel equally certain, have greatly shortened my life. I am therefore glad I did not get it, and I do not think I felt any disappointment at the time. Becoming tired of London and wishing for a country life, I took a small house at Barking in 1870, and in 1871 leased four acres of ground at Grays, including a very picturesque well-timbered old chalk-pit, above which I built a house having a very fine view across to the hills of North Kent and down a reach of the Thames to Gravesend.

Seven years later, in 1878, when Epping Forest had been acquired by the Corporation of London, a superintendent was to be appointed to see to its protection and improvement while preserving its "natural aspect" in accordance with the Act of Parliament which restored it to the public. This position would have suited me exactly, and if I had obtained it and had been allowed to utilize the large extent of open unwooded land in the way I suggested in my article in the *Fortnightly Review* ("Epping Forest, and how best to deal with it"), an experiment in illustration of the geographical distribution of plants would have been made which would have been both unique and educational, as well as generally interesting. I obtained recommendations and testimonials from the presidents of all the natural history societies in London, from numerous residents near the forest and in London, from many eminent men and members of Parliament—seventy in all; but the City merchants and tradesmen with whom the appointment lay wanted a "practical man" to carry out their own ideas, which were to utilize all the open spaces for games and sports, to build a large hotel close to Queen Elizabeth's hunting lodge, and to encourage excursions and school treats, allowing swings, round-abouts, and other such amusements.

My failure to obtain the post at Epping Forest was certainly a disappointment to me, but I am inclined to think now that even that was really for the best, since it left me free to do literary work which I should certainly not have done if I had had permanent employment so engrossing and interesting as that at Epping. In that case I should not have gone to lecture in America, and should not have written "Darwinism," perhaps none of my later books, and very few of the articles contained in my "Studies" This body of literary and popular scientific work is, perhaps, what I was best fitted to perform, and if so, neither I nor my readers have any reason to regret my failure to obtain the post of superintendent and guardian of Epping Forest.

CHAPTER XV

SIR CHARLES LYELL AND CHARLES DARWIN (1863–1881)

AMONG the eminent men of science with whom I became more or less intimate during the period of my residence in London, I give the first place to Sir Charles Lyell, not only on account of his great abilities and his position as one of the brightest ornaments of the nineteenth century, but because I saw more of him than of any other man at all approaching him as a thinker and leader in the world of science, while my correspondence with him was more varied in the subjects touched upon, and in some respects of more general interest, than my more extensive correspondence with Darwin.

I do not remember when I first saw Sir Charles Lyell, but I probably met him at some of the evening meetings of the scientific societies. I first lunched with him in the summer of 1863, and then met, for the first time, Lady Lyell and Miss Arabella B. Buckley. Miss Buckley had become Sir Charles's private secretary early in that year, and she informs me that she remembers this visit because Lady Lyell gave her impressions of me afterwards—I am afraid not very favourable ones, as I was shy, awkward, and quite unused to good society. With Sir Charles I soon felt at home, owing to his refined and gentle manners, his fund of quiet humour, and his intense love and extensive knowledge of natural science. His great liberality of thought and wide general interests were also attractive to me; and although when he had once arrived at a definite conclusion he held by it very tenaciously until a considerable body of well-ascertained facts could be adduced against it, yet he was always willing to listen to the arguments of his opponents, and to give them careful and repeated consideration.

In 1867, when a new edition of the "Principles of Geology" was in progress, I had much correspondence and many talks with Sir Charles, chiefly on questions relating to distribution and dispersal, in which he, like myself, was greatly interested. He was by nature so exceedingly cautious and conservative, and always gave such great weight to difficulties that occurred to himself or that were put forth by others, that it was not easy to satisfy him on any novel view upon which two opinions existed or were possible. We used often to discuss these various points, but in any case that seemed to him important he usually preferred to write to me, stating his objections, sometimes at great length, and asking me to give my views.

In the following year, when I was living at Hurst-pierpoint, I wrote a letter to Sir Charles on Darwin's new theory of "Pangenesis," a passage from which I will quote, because the disproof of it, which I thought would not be given, was not long in coming, and, with the more satisfactory theory of Weismann, led me to change my opinion entirely. I wrote (February 20, 1868): "I am reading Darwin's book ('Animals and Plants under Domestication'), and have read the 'Pangenesis' chapter first, for I could not wait. The hypothesis is *sublime* in its simplicity and the wonderful manner in which it explains the most mysterious of the phenomena of life. To me it is *satisfying* in the extreme. I feel I can never give it up, unless it be *positively* disproved, which is impossible, or replaced by one which better explains the facts, which is highly improbable. Darwin has here decidedly gone ahead of Spencer in generalization. I consider it the most wonderful thing he has given us, but it will not be generally appreciated."

This was written when I was fresh from the spell of this most ingenious hypothesis. Galton's experiments on blood transfusion with rabbits first staggered me, as it seemed to me to be the very disproof I had thought impossible. And later on, when Weismann adduced his views on the continuity of the germplasm, and the consequent non-heredity of acquired characters; and further, when he showed that the supposed transmission of such characters, which Darwin had accepted and which the hypothesis of pangenesis was constructed to account for, was not really proved by any evidence whatever;—I was compelled to discard Darwin's view in favour of that of Weismann, which is now almost everywhere accepted as being the most probable, as well as being the most in accordance with all the facts and phenomena of heredity.

The subject on which Sir Charles Lyell and myself had the longest discussions was that of the effects of the glacial period on the distribution of plants and animals, and on the origin of lake basins.

On the question of the ice-origin of Alpine lakes I could never get him to accept my extreme views. In March, 1869, I received from him a letter of thirteen pages, and another of thirty pages, on this and allied questions, setting forth the reasons why he rejected ice action as having ground out the larger lakes, much as he states them in the fourth edition of "The Antiquity of Man." At page 361 he says that "the gravest objection to the hypothesis of glacial erosion on a stupendous scale is afforded by the entire absence of lakes of the first magnitude in several areas where they ought to exist, if the enormous glaciers which once occupied those spaces had possessed the deep excavating power ascribed to them." He then goes on to adduce numerous places where he thinks there ought to have been lakes on the glacier theory, which are the same as he adduced in letters to myself, and which I answered in each case, and sometimes at great length, by similar arguments to those I have adduced in vol. i. chap.

v. of my "Studies, Scientific and Social." If any one who is interested in these questions, after considering Sir Charles Lyell's difficulties and objections in his "Antiquity of Man," will read the above chapter, giving special attention to the sections headed *The Conditions that favour the Production of Lakes by Ice-erosion,* and the following section on *Objections of Modern Writers considered,* I think he will, if he had paid any attention to the phenomena in glaciated regions, admit that I show the theory of ice-erosion to be the only one that explains all the facts.

In a letter written on April 28, 1869, after referring to Darwin's regret at the concluding passages of my *Quarterly Review* article on "Man," which he "would have thought written by some one else," I add the following summary on my position, perhaps more simply and forcibly stated than in any of my published works:—

"It seems to me that if we once admit the necessity of *any* action beyond 'natural selection' in developing man, we have no reason whatever for confining that agency to his brain. On the mere doctrine of chances it seems to me in the highest degree improbable that so many points of structure, all tending to favour his mental development, should concur in man alone of all animals. If the erect posture, the freedom of the anterior limbs from purposes of locomotion, the powerful and opposable thumb, the naked skin, the great symmetry of form, the perfect organs of speech, and, in his mental faculties, calculation of numbers, ideas of symmetry, of justice, of abstract reasoning, of the infinite, of a future state, and many others, cannot be shown to be each and all *useful* to man in the very lowest state of civilization— how are we to explain their co-existence in him alone of the whole series of organized being? Years ago I saw in London a bushman boy and girl, and the girl played very nicely on the piano. Blind Tom, the half-idiot negro slave, had a 'musical ear' or brain, superior, perhaps, to that of the best living musicians. Unless Darwin can show me how this latent musical faculty in the lowest races can have been developed through *survival* of the fittest, can have been of *use* to the individual or the race, so as to cause those who possessed it in a fractionally greater degree than others to win in the struggle for life, I must believe that some other power (than natural selection) caused that development. It seems to me that the *onus probandi* will lie with those who maintain that man, body and mind, could have been developed from a quadrumanous animal by 'natural selection.'"

In a letter to Darwin, written a week later and printed in the "Life, Letters, and Journals,"

Sir Charles quotes the preceding argument entire, and goes on to express his general agreement with it.

He then refers to the glacial-lake theory as follows:— "As to the scooping out of lake basins by glaciers, I have had a long, amicable, but controversial correspondence with

Wallace on that subject, and I cannot get over (as, indeed, I have admitted in print) an intimate connection between the number of lakes of modern date and the glaciation of the regions containing them. But as we do not know how ice can scoop out Lago Maggiore to a depth of 2600 feet, of which all but 600 is below the level of the sea, getting rid of the rock supposed to be worn away as if it was salt that had melted, I feel that it is a dangerous causation to admit in explanation of every cavity which we have to account for, including Lake Superior."

This passage shows, I think, that he was somewhat staggered by my arguments, but could not take so great a step without further consideration and examination of the evidence. I feel sure, therefore, that if he had had before him the numerous facts since made known, of erratic blocks carried by the ice to heights far above their place of origin in North America, and even in our own islands, as described at p. 75 and p. 90 of my "Studies" (vol. i.), with evidence of such action now occurring in Greenland (p. 91), of the Moel Tryfan beds having been forced up by the glacier that filled the Irish sea, he would have seen, I feel sure, that his objections were all answered by actual phenomena, and that the gradual erosion of Lago Maggiore was far within the powers of such enormous accumulations of ice as must have existed over its site.

During the ten years 1863–72, I saw a good deal of Sir

Charles. If he had any special subject on which he wished for information, he would sometimes walk across the park to St. Mark's Crescent for an hour's conversation; at other times he would ask me to lunch with him, either to meet some interesting visitor or for friendly talk. After my marriage my wife and I occasionally dined with him or went to his evening receptions. These latter were very interesting, both because they were not overcrowded and on account of the number of scientific and other men of eminence to be met there. Among these were Professor Tyndall, Sir Charles Wheatstone, Sir Charles Bunbury, Mr. Lecky, and a great many others. The Duke of Argyll was frequently there, and although we criticized each other's theories rather strongly, he was always very friendly, and we generally had some minutes' conversation whenever I met him. Miss Buckley (now Mrs. Fisher) was a very constant guest, and would point out to me the various celebrities who happened to be present, and thus began a cordial friendship which has continued unbroken, and has been a mutual pleasure and advantage. I therefore look back upon my friendship with Sir Charles Lyell with unalloyed satisfaction as one of the most instructive and enjoyable episodes in my life-experience.

Soon after I returned home, in the summer of 1862, Mr. Darwin invited me to come to Down for a night, where I had the great pleasure of seeing him in his quiet home, and in the midst of his family. A year or two later I spent a week-end with him in company with Bates, Jenner Weir, and a few other naturalists; but my most frequent interviews with him were when he spent a few weeks with his brother, Dr. Erasmus Darwin, in Queen Anne Street, which he usually did every year when he was well enough, in order to see his friends and collect information for his various works. On these occasions I usually lunched with him and his brother, and sometimes one other visitor, and had a little talk on some of the matters specially interesting him. He also sometimes called on me in St. Mark's Crescent for a quiet talk or to see some of my collections.

My first letter from him dealing with scientific matters was in August, 1862, and our correspondence was very extensive during the period occupied in writing or correcting his earlier books on evolution, down to the publication of "The Expression of the Emotions in Man and Animals," in 1872, and afterwards, at longer intervals, to within less than a year of his death. A considerable selection of our correspondence has been published in the "Life and Letters" (1887), and especially in "More Letters" (1903); while several of the more interesting of these were contained in the one-volume life, entitled "Charles Darwin," which appeared in 1892. As many of my readers, however, may not have these works to refer to, I will here give a few of his letters to myself which have not yet been published, and also occasional extracts from some that have already appeared, in order to make clear the nature of our discussions.

On February 23, 1867, he wrote to me asking if I could solve a difficulty for him. He says: "On Monday evening I called on Bates, and put a difficulty before him which he could not answer, and, as on some former similar occasion, his first suggestion was, 'You had better ask Wallace.' My difficulty is, Why are caterpillars sometimes so beautifully and artistically coloured? Seeing that many are coloured to escape dangers, I can hardly attribute their bright colour in other cases to mere physical conditions. Bates says the most gaudy caterpillar he ever saw in Amazonia was conspicuous at the distance of yards, from its black and red colours, whilst feeding on large, green leaves. If any one objected to male butterflies having been made beautiful by sexual selection, and asked why they should not have been made beautiful as well as their caterpillars, what would you answer? I could not answer, but should maintain my ground. Will you think over this, and some time, either by letter or when we meet, tell me what you think?"

On reading this letter, I almost at once saw what seemed to be a very easy and probable explanation of the facts. I had then just been preparing for publication (in the *Westminster Review*) my rather elaborate paper on "Mimicry and Protective Colouring," and the numerous cases in which specially showy and slow-flying butterflies were known to have a peculiar odour and taste which protected them from the attacks of insect-eating birds and other animals, led me at once to suppose that the gaudily coloured caterpillars must have a similar protection. I had just ascertained from Mr. Jenner Weir that one of our common white moths (*Spilosoma menthrasti*) would not be eaten by most of the small birds in his aviary, nor by young turkeys. Now, as a *white* moth is as conspicuous in the *dusk* as a *coloured* caterpillar in the *daylight,* this case seemed to me so much on a par with the other that I felt almost sure my explanation would turn out correct. I at once wrote to Mr. Darwin to this effect, and his reply, dated February 26, is as follows:—

"MY DEAR WALLACE,

Bates was quite right; you are the man to apply to in a difficulty. I never heard anything more ingenious than your suggestion, and I hope you may be able to prove it true. That is a splendid fact about the white moths; it warms one's very blood to see a theory thus almost proved to be true."

I then mentioned the subject at a meeting of the Entomological Society; and in the following year experiments were made by two gentlemen which proved clearly that while green or brown caterpillars were greedily eaten by all insect-eating birds, those which were gaudily coloured or were hairy, and usually exposed themselves on the plants on which they fed in the daytime, were rejected by almost all the birds and usually by lizards as well. The principle of "warning coloration," as it is termed, is now generally admitted as being widely prevalent in nature.

In the year 1870 Mr. A. W. Bennett read a paper before Section D of the British Association at Liverpool, entitled "The Theory of Natural Selection from a Mathematical Point of View," and this paper was printed in full in *Nature* of November 10, 1870. To this I replied on November 17, and my reply so pleased Mr. Darwin that he at once wrote to me as follows:—

"Down, November 22.

MY DEAR WALLACE,

I must ease myself by writing a few words to say how much I and all in this house admire your article in *Nature*. You are certainly an unparalleled master in lucidly stating a case and in arguing.

Nothing ever was better done than your argument about the term Origin of Species, and about much being gained if we know nothing about precise cause of each variation."

At the end of the letter he says something about the progress of his great work, "The Descent of Man."

"I have finished 1st vol. and am half-way through proofs of 2nd vol. of my confounded book, which half kills me by fatigue, and which I fear will quite kill me in your good estimation.

If you have leisure, I should much like a little news of

you and your doings and your family,

Ever yours very sincerely,

CH. DARWIN."

The above remark, "kill me in your good estimation," refers to his views on the mental and moral nature of man being very different from mine, this being the first important question as to which our views had diverged. But I never had the slightest feeling of the kind he supposed, looking upon the difference as one which did not at all affect our general agreement, and also as being one on which no one could dogmatize, there being much to be said on both sides. The last paragraph shows the extreme interest he took in the personal affairs of all his friends.

Soon after the "Descent of Man" appeared, I wrote to Darwin, giving my impressions of the first volume, to which he replied (January 30, 1871). This letter is given in the "Life and Letters" (iii. p. 134), but I will quote two short passages expressing his kind feelings towards myself. He begins, "Your note has given me very great pleasure, chiefly because

I was so anxious not to treat you with the least disrespect, and it is so difficult to speak fairly when differing from any one. If I had offended you, it would have grieved me more than you will readily believe." And the conclusion is, "Forgive me for scribbling at such length. You have put me quite in good spirits; I did so dread having been unintentionally unfair towards your views. I hope earnestly the second volume will escape as well. I care now very little what others say. As for our not agreeing, really, in such complex subjects, it is almost impossible for two men who arrive independently at their conclusions to agree fully; it would be unnatural for them to do so."

Again, on July 12, he writes: "I feel very doubtful how far I shall succeed in answering Mivart. It is so difficult to answer objections to doubtful points and make the discussion readable. The worst of it is, that I cannot possibly hunt through all my references for isolated points—it would take me three weeks of intolerably hard work. I wish I had your power of arguing clearly. At present I feel sick of everything, and if I could occupy my time and forget my daily discomforts, or rather miseries, I would never publish another word. But I shall cheer up, I dare say, soon, having only just got over a bad attack. Farewell. God knows why I bother you about myself.

"I can say nothing more about missing links than I have said. I should rely much on pre-Silurian times; but then comes Sir W. Thomson like an odious spectre. Farewell."

This last remark refers to the limitation of the earth's age by physicists, so as not to leave time enough for the evolution of organisms.

During this latter period of his life I had but little correspondence with him, as I had no knowledge whatever of the subjects he was then working on. But he still continued to write to me occasionally, either referring kindly to my own work or sending me facts or suggestions which he thought would be of interest to me. I will here give only some extracts from a few of the latest of the letters I received from him.

On November 3, 1880, he wrote me the following very kind letter upon my "Island Life," on which I had asked for his criticism:—

"I have now read your book, and it has interested me deeply. It is quite excellent, and seems to me the best book which you have ever published; but this may be merely because I have read it last. As I went on I made a few notes, chiefly where I differed slightly from you; but God knows whether they are worth your reading. You will be disappointed with many of them; but it will show that I had the will, though I did not know the way to do what you wanted.

I have said nothing on the infinitely many passages and views, which I admired and which were new to me. My notes are badly expressed, but I thought that you would excuse my taking any pains with my style. I wish my confounded handwriting was better. I had a note the other day from Hooker, and I can see that he is much pleased with the dedication."

With this came seven foolscap pages of notes, many giving facts from his extensive reading which I had not seen.

In another letter, two months later, he recurs to the same subject.

"Down, January 2, 1881.

MY DEAR WALLACE,

The case which you give is a very striking one, and I had overlooked it in *Nature*;[1] but I remain as great a heretic as ever. Any supposition seems to me more probable than that the seeds of plants should have been blown from the mountains of Abyssinia, or other central mountains of Africa, to the mountains of Madagascar. It seems to me almost infinitely more probable that Madagascar extended far to the south during the glacial period, and that the S. hemisphere was, according to Croll, then more temperate; and that the whole of Africa was then peopled with some temperate forms, which crossed chiefly by agency of birds and sea-currents, and some few by the wind, from the shores of Africa to Madagascar subsequently ascending to the mountains.

How lamentable it is that two men should take such widely different views, with the same facts before them; but this seems to be almost regularly our case, and much do I regret it. I am fairly well, but always feel half dead with fatigue. I heard but an indifferent account of your health some time ago, but trust that you are now somewhat stronger.

Believe me, my dear Wallace,

Yours very sincerely,

CH. DARWIN."

¹ *Nature,* December 9, 1880. The substance of this article by Mr. Baker, of Kew, is given in "More Letters," vol. iii. p. 25, in a footnote.

It is really quite pathetic how much he felt difference of opinion from his friends. I, of course, should have liked to be able to convert him to my views, but I did not feel it so much as he seemed to do. In letters to Sir Joseph Hooker (in February and August, 1881) he again states his view as against mine very strongly ("More Letters," iii. pp. 25 and 27); and this, so far as I know, is the last reference he made to the subject. The last letter I received from him was entirely on literary and political subjects, and, as usual, very kind and friendly. As it makes no reference to our controversies, and touches on questions never introduced before in our correspondence, I think it will be interesting to give it entire.

"Down, July 12, 1881.

MY DEAR WALLACE,

I have been heartily glad to get your note and hear some news of you. I will certainly order 'Progress and Poverty,' for the subject is a most interesting one. But I read many years ago some books on political economy, and they produced a disastrous effect on my mind, viz., utterly to distrust my own judgment on the subject, and to doubt much every one else's judgment! So I feel pretty sure that Mr. George's book will only make my mind worse confounded than it is at present. I also have just finished a book which has interested me greatly, but whether it would interest any one else I know not. It is the 'Creed of Science,' by W. Graham, A.M. Who or what he is I know not, but he discusses many great subjects, such as the existence of God, immortality, the moral sense, the progress of society, etc. I think some of his propositions rest on very uncertain foundations, and I could get no clear idea of his notions about God. Notwithstanding this and other blemishes, the book has interested me *extremely*. Perhaps I have been to some extent deluded, as he manifestly ranks too high what I have done.

I am delighted to hear that you spend so much time out-of-doors and in your garden. From Newman's old book (I forget title) about the country near Godalming, it must be charming.

We have just returned home after spending five weeks on Ullswater. The scenery is quite charming, but I cannot walk, and everything tires me, even seeing scenery, talking with any one, or reading much. What I shall do with my few remaining years of life I can hardly tell. I have everything to make me happy and contented, but life has become very wearysome to me. I heard lately from Miss Buckley in relation to Lyell's Life, and she mentioned that you were thinking of Switzerland, which I should think and hope that you would enjoy much.

I see that you are going to write on the most difficult political question, the land. Something ought to be done, but what is the rub. I hope that you will (not) turn renegade to natural history; but I suppose that politics are very tempting.

With all good wishes for yourself and family,

Believe me, my dear Wallace,

Yours very sincerely,

CHARLES DARWIN."

This letter is, to me, perhaps the most interesting I ever received from Darwin, since it shows that it was only the engrossing interests of his scientific and literary work, performed under the drawback of almost constant ill-health, that prevented him from taking a more active part in the discussion of those social and political questions that so deeply affect the lives and happiness of the great bulk of the people. It is a great satisfaction that his last letter to me, written within nine months of his death, and terminating a correspondence which had extended over a quarter of a century, should be so cordial, so sympathetic, and broad-minded.

In 1870 he had written to me, "I hope it is a satisfaction

to you to reflect—and very few things in my life have been more satisfactory to me—that we have never felt any jealousy towards each other, though in some sense rivals. I believe I can say this of myself with truth, and I am absolutely sure that it is true of you." This friendly feeling was retained by him to the last, and to have thus inspired and retained it, notwithstanding our many differences of opinion, I feel to be one of the greatest honours of my life.

In conclusion, it may interest my readers if I give briefly the four chief points on which I differed from Darwin. They are as follows: —

(1) Darwin held that man had been developed physically and intellectually by continuous modification under natural selection from some ancestral form, whilst I, though agreeing with him with regard to man's physical form, believed that some agency other than natural selection, and analogous to that which first produced organic life, had brought into being his moral and intellectual qualities.

(2) Darwin believed that in the case of certain animals the males had obtained their bright colours or other ornaments by selection through female choice. I, on the other hand, believed that natural selection had operated independently on the two sexes, and each had acquired coloration or form according to its need for protection. The females, being often more exposed to danger than the males (as in the case of sitting birds), had acquired more subdued coloration

whilst the males had remained bright and comparatively conspicuous.

(3) Darwin thought that the arctic plants found on isolated mountain tops within the tropics could only be explained by the spreading of the arctic flora over the tropics during the glacial period. From a study of the floras of oceanic islands, I had come to the conclusion that the mountain flora had been derived by aeumlrial transmission of seeds either by birds or by gales.

(4) Darwin always believed in the inheritance of acquired characteristics, such as the results of use or disuse of organs, and the effects of climate, food, etc., on the individual. I also accepted this theory at first, but when I had studied Mr. Galton's experiments and Dr. Weismann's theory of the continuity of the germplasm I had to change my views. The latter theory really simplifies and strengthens the fundamental doctrine of natural selection.

It will thus appear that none of my differences of opinion from Darwin imply any real divergence as to the overwhelming importance of the great principle of natural selection, while in several directions I believe that I have extended and strengthened it. The principle of "utility," which is one of its chief foundation-stones, I have always advocated unreservedly; while in extending this principle to almost every kind and degree of coloration, and in maintaining the power of natural selection to increase the infertility of hybrid unions, I have considerably extended its range. Hence it is that some of my critics declare that I am more Darwinian than Darwin himself, and in this, I admit, they are not far wrong.

When Darwin died in 1881 I was honoured by an invitation to his funeral in Westminster Abbey, as one of the pall-bearers, along with nine of his most distinguished friends or admirers, among whom was J. Russell Lowell, as the representative of American science and literature. Among the many obituary notices of Darwin, that by Huxley (in *Nature*, of April 27) is one of the shortest, most discriminating, and most beautiful. It is published also in the second volume of his "Collected Essays." For those who have not read this true and charming estimate of his friend, I may quote one passage: "One could not converse with Darwin without being reminded of Socrates. There was the same desire to find some one wiser than himself; the same belief in the sovereignty of reason; the same ready humour; the same sympathetic interest in all the ways and works of men. But instead of turning away from the problems of nature as wholly insoluble, our modern philosopher devoted his whole life to attacking them in the spirit of Heraclitus and Democritus, with results which are as the substance of which their speculations were anticipatory shadows."

CHAPTER XVI

HERBERT SPENCER, HUXLEY, AND OTHER FRIENDS

SOON after my return home, in 1862, Bates and I, having both read "First Principles" and been immensely impressed by it, went together to call on Herbert Spencer, I think by appointment. Our thoughts were full of the great unsolved problem of the origin of life—a problem which Darwin's "Origin of Species" left in as much obscurity as ever—and we looked to Spencer as the one man living who could give us some clue to it. His wonderful exposition of the fundamental laws and conditions, actions and interactions of the material universe seemed to penetrate so deeply into that "nature of things" after which the early philosophers searched in vain, and whose blind gropings are so finely expressed in the grand poem of Lucretius, that we both hoped he could throw some light on that great problem of problems. He was very pleasant, spoke appreciatively of what we had both done for the practical exposition of evolution, and hoped we would continue to work at the subject. But when we ventured to touch upon the great problem, and whether he had arrived at even one of the first steps towards its solution, our hopes were dashed at once. That, he said, was too fundamental a problem to even think of solving at present. We did not yet know enough of matter in its essential constitution nor of the various forces of nature; and all he could say was that everything pointed to life having been a development out of matter—a phase of that continuous process of evolution by which the whole universe had been brought to its present condition. So we had to wait and work contentedly at minor problems. And now, after forty years, though Spencer and Darwin and Weis-mann have thrown floods of light on the phenomena of life, its essential nature and its origin remain as great a mystery as ever. Whatever light we do possess is from a source which Spencer and Darwin neglected or ignored.

The first letter I received from Spencer was when I sent him my paper on "The Origin of Human Races under the Law of Natural Selection." He said that he had read it with great interest, and added, "Its leading idea is, I think, undoubtedly true," concluding with a hope that I would pursue the inquiry.

Soon afterwards he invited me to dine with him in Bayswater, where he lived for many years in a boarding-house with rather a commonplace set of people; and I visited him there several times.

In 1872, in my presidential address to the Entomological Society, I endeavoured to expound Herbert Spencer's theory of the origin of insects, on the view that they are fundamentally *compound animals,* each segment representing one of the original independent organisms. On sending him a copy of the address, he wrote to me as follows: "It is gratifying to me to find that your extended knowledge does not lead you to scepticism respecting the speculation of mine which you quote, but rather enables you to cite further facts in justification of it. Possibly your exposition will lead some of those, in whose lines of investigation the question lies, to give deliberate attention to it."

In 1874, when writing "The Principles of Sociology," Herbert Spencer asked me to look over the proofs of the first six chapters, and give him the benefit of my criticisms, "alike as naturalist, anthropologist, and traveller." I found very little indeed requiring emendation, but I sent him a couple of pages of notes with suggestions on points of detail, which, I believe, were of some use to him.

During the year 1881 I had several letters from him, dealing with subjects of general interest. In consequence of an article I wrote on "How to Nationalize the Land," especially showing how to avoid the supposed insuperable objection of State management, a "Land Nationalization Society" was formed, of which I was chosen president. As I had been induced to study the question by Herbert Spencer's early volume on "Social Statics," I sent him a copy of our programme and asked if he would join us. His reply is very instructive, as showing how nearly he agreed with us at that time, and also how slight were the difficulties he suggested as the most important.

The letter is as follows:—

"38, Queen's Gardens, Bayswater, W.,
April 25, 1881.
DEAR MR. WALLACE,

As you may suppose, I fully sympathize in the general aims of your proposed Land Nationalization Society; but for sundry reasons I hesitate to commit myself, at the present stage of the question, to a programme so definite as that which you send me. It seems to me that before formulating the idea in a specific shape, it is needful to generate a body of public opinion on the general issue, and that it must be some time before there can be produced such recognition of the general principle involved as is needful before definite plans can be set forth to any purpose.

It seems to me that the thing to be done at present is to arouse public attention to (1) the abstract inequity of the present condition of things; (2) to show that even now there is in our law a tacit denial of absolute private ownership, since the State reserves the power of resuming possession of land on making compensation; (3) that this tacitly admitted ownership ought to be overtly asserted; (4) and that having been overtly asserted, the landowner should be distinctly placed in the position of a tenant of the State on something like the terms proposed in your scheme: namely, that while the land itself should be regarded as public property, such value as has been given to it should vest in the existing so-called owner.

The question is surrounded with such difficulties that I fear anything like a specific scheme for resumption by the State will tend, by the objections made, to prevent recognition of a general truth which might otherwise be admitted. For example, in definitely making the proposed distinction between 'inherent value as dependent on natural conditions, etc.,' and the 'increased value given by the owner,' there is raised the questions—How are the two to be distinguished? How far back are we to go in taking account of the labour and money expended in giving fertility? In respect of newly enclosed tracts, some estimation may be made; but in respect of the greater part, long reduced to cultivation, I do not see how the valuations, differing in all cases, are to be made.

I name this as one point; and there are many others in respect of which I do not see my way. It appears to me that at present we are far off from the time at which action may advantageously be taken.

Truly yours,

HERBERT SPENCER."

The last three letters I received from Herbert Spencer were in 1894 and 1895, all on the subject of what he termed "the absurdity of Lord Salisbury's representation of the process of natural selection" in his British Association address at Oxford, wishing me to write to the *Times,* pointing out his errors, which were influencing many persons and writers in the press, and suggesting certain points I should especially deal with. He concluded, "It behoves you of all men to take up the gauntlet he has thus thrown down." I replied, declining the task, on the ground that I did not think Lord Salisbury's influence in a matter of science of much importance, and that I thought my time better employed in writing such articles on social and political, as well as general scientific questions which then interested me. To this he replied that he did not at all agree with me, and that "articles in the papers show that Lord Salisbury's argument is received with triumph, and unless it is disposed of, it will lead to a public reaction against the doctrine of evolution at large."

As I still declined to go into this controversy, having dealt with the whole matter in my "Darwinism," and still being sceptical as to any great effects being produced by the address in question, he wrote me a month later as follows: "As I cannot get you to deal with Lord Salisbury, I have decided to do it myself, having been finally exasperated into doing it by this honour paid to his address in France —the presentation of a translation to the French Academy.

The impression produced upon some millions of people in England cannot be allowed to be thus further confirmed without protest." He then asked me for some references, which I sent him, and his criticism of Lord Salisbury duly appeared, and was thoroughly well done, so that I had no reason to regret not having undertaken it myself. This was the latest letter I received from him; but during his last illness my wife, being in Brighton, called to make inquiries after his health, and left our cards, and I received a kindly expressed card in reply, written by his amanuensis, but signed with his own initials. It is dated November 28, 1903, ten days before his death.

Among his intimate friends, Herbert Spencer was always interesting from the often unexpected way in which he would apply the principles of evolution to the commonest topics of conversation, and he was always ready to take part in any social amusement. He once or twice honoured me by coming to informal meetings of friends at my little house in St. Mark's Crescent, and I also met him at Sir John Lubbock's very pleasant week-end visits, and also at Huxley's, in St. John's Wood. Once I remember dining informally with Huxley, the only other guests being Tyndall and Herbert Spencer. The latter appeared in a dress-coat, whereupon Huxley and Tyndall chaffed him, as setting a bad example, and of being untrue to his principles, quoting his Essay on "Manners and Fashion," but all with the most good-humoured banter.

Spencer took it in good part, and defended himself well, declaring that the coat was a relic of his early unregenerate days, and where could he wear it out if not at the houses of his best friends? "Besides," he concluded, "you will please to observe that I *am* true to principle in that I do *not* wear a white tie!"

Those who are acquainted only with the volumes of Herbert Spencer's "Synthetic Philosophy" can have no idea of the lightness, the energy, and the bright satire of some of his more popular writings. Such are many of his earlier Essays, and in his volume on "The Study of Sociology" we find abundant examples of these qualities.

With the remainder of my scientific friends I had, for the most part, only social intercourse, with no correspondence of general interest. Those I saw most of during my residence in London, and with whom I became most intimate, were Huxley, Tyndall, Sir John Lubbock, Dr. W. B. Carpenter, Sir William Crookes, Sir Joseph Hooker, Mr. Francis Galton, Professor Alfred Newton, Dr. P. L. Sclater, Mr. St. George Mivart, Sir William Flower, Sir Norman Lockyer, Professor R. Meldola, and many others whose names are only known to specialists. All these I met very frequently at scientific meetings, or at some of their houses at which I was occasionally a guest. To all of them I have been more or less indebted for valuable information or useful suggestions in the course of my work, and to these I must also add Professor E. B.

Poulton, of Oxford; F. W. H. Myers, Professors W. F. Barrett and Percival Wright, of Dublin, with Professors Patrick Geddes, of Edinburgh, and J. A. Thomson, of Aberdeen. For the last quarter of a century I have lived so completely in the country that I have ceased to have personal intercourse with most of them; and of those still among us, I can only say here that I hope and believe they all continue to be my very good friends. I may have to refer to some of them again, in connection with special conditions of my life. Here I will only give a few indications as to my personal relations with some of them.

Of all those I have mentioned I became, I think, most intimate with Huxley. At an early date after my return home he asked me to his house in Marlborough Place, where I soon became very friendly with his children, then all quite young, all very animated, and not at all shy. Mrs. Huxley was also exceedingly kind and pleasant, and the whole domestic tone of the house was such as to make me quite at my ease, which happens to me with only a few persons. I used often to go there on Sunday afternoons, or to spend the evening, while I was several times asked to dine to meet persons of similar pursuits to my own.

I often called in at Jermyn Street if I had any question to ask Huxley, and he was always ready to give me all the information in his power; while I am pretty sure I owe partly, if not largely, to his influence the grant of the royal medal of the Royal Society, and perhaps also of the Darwin medal. Once only there was a partial disturbance of our friendly relations, of the exact cause of which I have no record or recollection. I had published some paper in which, I believe, I had stated some view which he had originated without mentioning his name, and in such a way as to leave the impression that I put it forth as original. This I had no notion of doing; but I think it was an idea which had become so familiar to me that I had quite forgotten *who* originated it. I fancy some one must have called Huxley's attention to it, and when I next met him, I think just as he was leaving Jermyn Street to go home, he was much put out, and said something intimating that after what I had said in this paper, he wondered at my speaking to him again. I forget what more was said, but on going home I looked at the article, and found that I had used some expression that *might* be interpreted as a slight to him. I immediately wrote a letter of explanation and regret, and I here give his reply, which greatly relieved me, and our relations at once resumed their usual friendly character.

"MY DEAR WALLACE,

Very many thanks for your kind letter.

I am exceedingly callous to the proceedings of my enemies, but (I suppose by way of compensation) I am very sensitive to those of valued friends, and I certainly felt rather sore when I read your paper. But I dare say I should have 'consumed my own smoke' in that matter as I do in most, if I had not been very tired, very hungry, very cold, and consequently very irritable, when I met you yesterday. Pray forgive me if I was too plain spoken,

And believe me, as always,

Yours very faithfully,

T. H. Huxley.

Jermyn Street, February 14, 1870."

Although Huxley was as kind and genial a friend and companion as Darwin himself, and I was quite at ease with him in his family circle, or in after-dinner talk with a few of his intimates, and although he was two years younger than myself, yet I never got over a feeling of awe and inferiority when discussing with him any problem in evolution or allied subjects—an inferiority which I did not feel either with Darwin or Sir Charles Lyell. This was due, I think, to the fact that an enormous amount of Huxley's knowledge was of a kind of which I possessed only an irreducible minimum, and in which I often felt my deficiency. In the general anatomy and physiology of the whole animal kingdom, living and extinct, Huxley was a master, the equal—perhaps the superior—of the greatest authorities on these subjects in the scientific world; whereas I had never had an hour's instruction in either of them, had never seen a dissection of any kind, and never had any inclination to practise the art myself. Whenever I had to touch upon these subjects, or to use them to enforce my arguments, I had to get both my facts and my arguments at second hand, and appeal to authority both for facts and conclusions from them. And because I was thus ignorant, and because I had a positive distaste for all forms of anatomical and physiological experiment, I perhaps overestimated this branch of knowledge and looked up to those who possessed it in a pre-eminent degree as altogether above myself.

With Darwin and Lyell, on the other hand, although both possessed stores of knowledge far beyond my own, yet I did possess *some* knowledge of the same kind, and felt myself in a position to make use of their facts and those of all other students in the same fields of research quite as well as the majority of those who had observed and recorded them. I had, however, very early in life noticed, that men with immense *knowledge* did not always know how to draw just conclusions from that knowledge, and that I myself was quite able to detect their errors of reasoning. I have never hesitated to differ from Lyell, Darwin, and even Spencer, and, so far as I can judge, in all the cases in which I have so differed, the weight of scientific opinion is gradually turning in my direction. In reasoning power upon the general phenomena of nature or of society, I feel able to hold my own with them; my inferiority consists in my limited knowledge, and perhaps also in my smaller power of concentration for long periods of time.

With Huxley also I felt quite on an equality when dealing with problems arising out of facts equally well known to both of us; but wherever the structure or functions of animals were concerned, he had the command of a body of facts so extensive and so complex that no one who had not devoted years to their practical study could safely attempt to make use of them. I therefore never ventured to infringe in any way on his special departments of study, though I occasionally make use of some of the results which he so lucidly explained.

Among the more prominent naturalists, one of my chief friends, and the one whose society I most enjoyed, was Professor St. George Mivart, who for some years lived not far from us in London. He was a rather singular compound mentally, inasmuch as he was a sincere but thoroughly liberal Catholic, and an anti-Darwinian evolutionist. But his friendly geniality, his refined manners, his interesting conversation and fund of anecdote of the most varied kind, rendered him a charming companion. His most intimate friends seemed to be priests, one or two of whom were almost always among the guests, and often the only ones, when I dined with him. And they, too, were excellent company, full of humour and anecdote of the most varied kind, though also ready for serious talk or discussion; but in either case, with none of the reserve or somewhat rigid decorum of the majority of our clergy.

Mivart was a very severe and often an unfair critic of Darwin, and I never concealed my opinion that he was not justified in going so far as he did. I also criticized some of his own writings, but he took it all very good-naturedly, and we always remained excellent friends. Besides natural history we had other tastes in common. He enjoyed country life, and for some time had a small country house in the wilds of Sussex, about midway between Forest Row and Hayward's Heath, where we sometimes spent a few days. He was also greatly interested in psychical research and spiritualistic phenomena; but this I shall refer to again when I come to my own experiences and inquiries on this intensely interesting subject.

About a year or two after I had returned home, Sir James Brooke had also returned to England, and had retired to a small estate at the foot of Dartmoor, where he lived in a comfortable cottage-farmhouse amid the wild scenery in which he delighted. I had met him once or twice in London, and, I think in the summer of 1863 or 1864, he invited me to spend a week with him in Devonshire, to meet his former private secretary and my old friend in Sarawak, Mr. (now Sir Spencer) St. John. We had a very pleasant time, strolling about the district or taking rides over Dartmoor; while at meals we had old-time events to talk over, with discussions of all kinds of political and social problems in the evening. At the same time Lady Burdett-Coutts, with her friend Mrs. Brown, were staying near, and often drove over and took us all for some more distant excursions.

This meeting and my friendship with Sir James Brooke led to my receiving several invitations to dine in Stratton Street, where my friend George Silk was also a frequent guest; but my unfortunate habit of speaking my thoughts too plainly broke off the acquaintance. The rajah's nephew, Captain Brooke, who had been formerly designated as Sir James's successor under the Malay title of Tuan Muda (young lord), had done or written something (I forget what) to which Sir James objected, and a disagreement ensued, which resulted in the captain being deposed from the heirship, and his younger brother Charles, the present rajah, being nominated instead. As I was equally friendly and intimate with both parties and heard both sides, I thought the captain had been rather hardly treated, and one day, when the subject was mentioned at Stratton Street, I ventured to say so. This evidently displeased Lady Burdett-Coutts, and I was never invited again—a matter which did not at all disturb me, as the people I met there were not very interesting to me. When Sir James Brooke heard of my indiscretion, he wrote to me very kindly, saying that he knew that I was the captain's friend and had a perfect right to take his part, and that my doing so did not in the least offend him and would make no difference in our relations, and I continued to receive friendly letters from him till he went to Borneo for the last time, in 1866. Soon after his return he died at his Devonshire home, in June, 1868. I have given my estimate of his character and of his beneficent work at Sarawak in my "Malay Archipelago."

Among the dearest of my friends, the one towards whom I felt more like a brother than to any other person, was Dr. Richard Spruce, one of the most cultivated and charming of men, as well as one of the most enthusiastic and observing of botanists. As he lived in Yorkshire after 1867, I only saw him at rather long intervals, but I generally took the opportunity of lecture engagements in the north to pay him a few days' visit. Our correspondence also was scanty, as he was a great invalid and could not write much, and I only preserved such letters as touched upon subjects connected with my own work.

One of the most interesting, amusing, and eccentric men I became acquainted with during my residence in London, and with whom I soon became quite intimate, was Dr. T. Purland, a dentist, living in Mortimer Street, Cavendish Square. He was a stout, dark, middle-aged man, with somewhat Jewish features, and of immense energy and vitality—one of those men whose words pour out in a torrent, and who have always something wise or witty to say. He had been a great coin-collector, and had many anecdotes to tell of rarities hit upon accidentally. He had an unbounded admiration for Greek coins as works of art, and would dilate upon their beauties as compared with the poor and inartistic works of our day. He was something of an Egyptologist, and had many odds and ends of antiquities, including teeth from mummies and dentists' instruments found in the old tombs and sarcophagi. He was a widower with three growing-up children, and had been obliged to part with all the more valuable parts of his collection to educate them.

He was a very powerful mesmerist, and helped, with Dr. Elliotson and others, in establishing the mesmeric hospital then in existence, and could succeed in sending patients into the mesmeric trance when other operators failed. H was one of the few men at that time who had been up in a balloon (with Green, the celebrated aëronaut, I think), and one evening at our house in St. Mark's Crescent, when Huxley and Tyndall were present, he made some remarks which interested Tyndall, who thereupon asked him many questions as to his sensations, the general appearance of the earth, clouds, etc., to all of which Dr. Purland replied with such promptitude and intelligence that all our friends were soon gathered round to hear the discussion, which went on a long time.

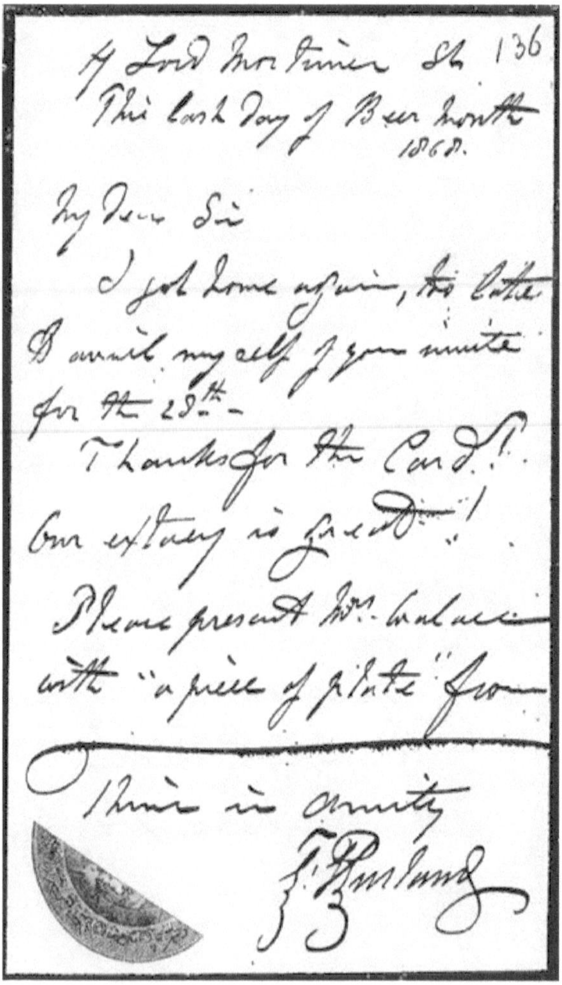

4 Lord Mortimer St. 136

The last day of Beer month

1868.

My Dear Sir

I got home again, & I rather

B avail myself of your invite

for the 28th –

Thanks for the Card!

Our ecstasy is great!

Please present Mrs. Wallace

with "a piece of plate" from

Thine in Amity

F. Rowland

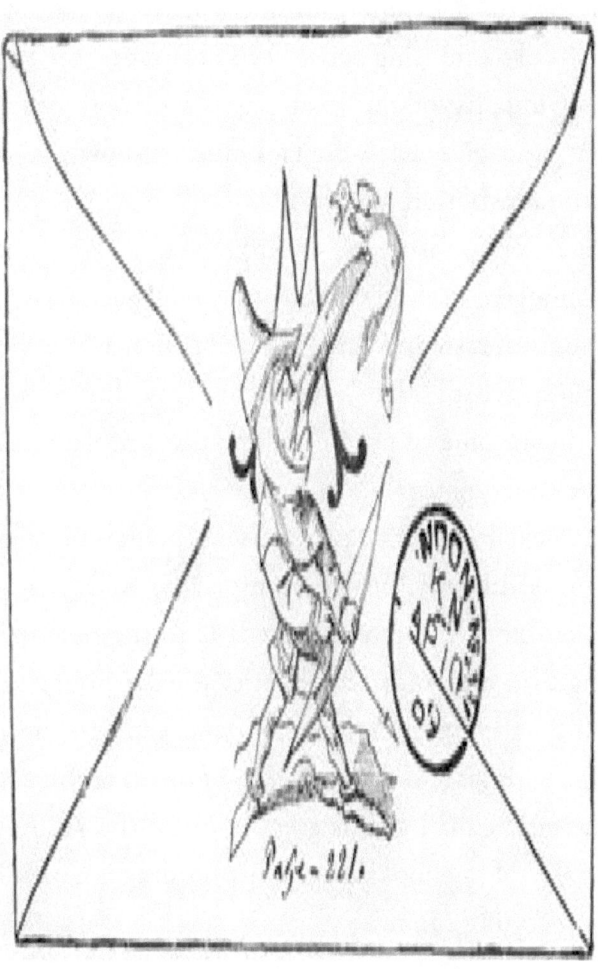

I possess about a dozen of his letters—replies to invitations, remarks on reading my early books, or other matters—all so amusing and so well illustrating the character and individuality of the man that I will now print some of them, and give a few in facsimile to show his style of caricature illustration.

The letter opposite was, I think, the first I had from him, and I only give it to illustrate two of his peculiarities—his gastronomical taste indicated by "Beer Month" for October, and the "piece of plate" represented by *half* a beautiful little print in blue of an old willow-pattern plate pasted in opposite the signature.

The next letter is in answer to an invitation to tea. He had been reading my "Malay Archipelago," and the reference on the envelope (here reproduced) is to the description of the king bird of paradise. The letter itself is in hieroglyphic form, intimating that he had other engagements, indicating himself by his large nose and scrap-book under his arm.

The next I shall give is an account of the sad results of reading one of my books aloud. The heading is a pseudonym for his operating room.

"Fang Castle, June, 1870,
Therm. 77¼.

Thanks worthy Signor for the entertainment afforded by your Boke on *Natural Selection*. But good as 'Natural

Selection' is, or maybe: I like *Mutual Selection* much better; and to my thinking it is of much more importance: ex. gr. mutual selection is this—A Lady asks me to become her husband I ax her to become my wife—that's *Mutual Selection*—aint it 'Natural'? The question of the 'fittest' is a subsequent affair: as is the *Creation by birth,* etc., etc.

But the pleasure was sadly and suddenly interrupted: I was reading aloud, and got on pretty well through p. 90–91. At 92 Jaws ached terribly! but at p. 94 and 5, even vulcanite could not stand it; and to my horror my upper set of teeth gave way with a crash! divided between the right lateral and the canine. I was helpless; and but for an old piece in reserve, my enjoyment of a succulent Roast Pig would have been entirely destroyed: it cost me dear —quite the value of a collection: I must give up reading *scientific (?) names* aloud.

I picked up a good specimen of Lignum ambulans for a shilling a week ago: and it now forms a prominent feature in our surgery. We are promised a Phyllium in a few days: and a Kallima paralekta. The Rosa Canina is a puzzle at present: I never saw a Red Canine tooth! Speaking of teeth—Huxley in his Physiology says Bicuspids *never* have more than two fangs—He knows nothing about it. I have them with three— Molars with 4, 5, and 6! In my lecture case, now before me, there are several: they are not as common as dirt or *earwigs in the country!* but they often turn up.

I begin the second reading to-night—not aloud—*oh no!*
With our best Salaam to the Lady, I remain
Thine in amity,
THEODOSIUS PURLAND."

The last I have was an anecdote of animal sagacity, a subject then being discussed in the papers, and of which he had given me some examples. I give a print of it, as it is a good example of his caricature drawing and of one of his fantastic signatures.

Our pleasant intimacy came to an end in a most absurd manner. Dr. Purland was, as I have said, a powerful and enthusiastic mesmerist, and had given his services for many surgical operations. Just as the opposition of the chiefs of the medical profession was dying away, and they were beginning to acknowledge the great value of the mesmeric sleep in alleviating pain and greatly facilitating serious operations, the discovery of anæsthetics offered a rival, which, though much more dangerous, was more certain and more easily applied in emergencies, and this led to the discontinuance of the use of mesmerism as a remedial agent. This naturally disgusted Dr. Purland, who, with the whole energy of his character, hated chloroform, ether, and nitrous-oxide gas, and would have nothing to do with them in his profession. Besides, he despised any one who could not bear the pain of tooth-drawing, and would turn away any patient who required the gas to be administered. A year or two after the date of his last letter I had occasion to approach him in his professional capacity. Entirely forgetting his objections, which, in fact, I had hardly believed to be real, after making an appointment I asked him to get a doctor to administer nitrous-oxide, as I could not stand the pain of three or four extractions in succession. This thoroughly enraged him. He wrote me a most violent letter, saying he could not continue to be the friend of a man who could ask him to do such a thing, and gave me the name of an acquaintance of his who had no such scruples and whose work was thoroughly good. And that was the last communication I ever had from Dr. Purland.

I will conclude this chapter with a few words about those meetings of the British Association at which I was present. In 1862 I was invited by my kind friend, Professor Alfred Newton, to be his guest at Magdalen College during the meeting, in company with a party of scientific friends, chiefly ornithologists. This was my first visit both to Cambridge and to the Association, and under such pleasant conditions I thoroughly enjoyed both. Besides the number of eminent men of science I had the opportunity of hearing or seeing, I had the pleasure of spending an evening with Charles Kingsley in his own house, and enjoying his stimulating conversation. There was also a slight recrudescence of the evolution controversy in the rather painful dispute between Professor Richard Owen and Huxley, supported by Flower, on certain alleged differences between the brains of man and apes.

I so much enjoyed the meeting, both in its scientific and social aspects, that I attended the next eleven meetings, and generally took part in some of the discussions, besides occasionally reading short papers. One of the most enjoyable meetings socially was that at Exeter, where I and a large party of scientific men were hospitably entertained at a country mansion eight or ten miles from the city, into which we were driven and brought back every day. Among the guests there was Professor Rankin, who entertained us by singing some of his own descriptive or witty compositions, especially the "Song of the Engine Driver," and that inimitable Irish descriptive song on "The City of Mullingar." On this occasion there appeared one of the most humorous parodies of the work of the Association that has ever been written, called "Exeter Change for British Lions." It was in the form of a small magazine, giving reports of the meetings, with absurd papers, witty verses, and clever parodies of the leading members, all worthy of Hood himself in his most humorous vein.

I was also honoured by being admitted to the fraternity of the "Red Lions," who fed together during each meeting of the Association and expressed applause by gentle roars and wagging of (coat) tails. On these occasions all kinds of jokes were permissible, and speeches were made and songs sung by the scientific humourists assembled. At Edinburgh, in 1871, Lord Neaves, a well-known wit and song-writer, was a guest, and gave us some of his own compositions, especially that on "The Origin of Species à la Darwin"—which he recited standing up and with very fine humour. The following verses are samples:—

"A *very* tall Pig with a *very* long nose
Sends forth a proboscis right down to his toes,
And then by the name of an Elephant goes,

Which Nobody can Deny!

"An Ape with a pliable thumb and big brain,
When the gift of the gab he had managed to gain,
As Lord of Creation established his reign,

Which Nobody can Deny!"

And so on for twelve verses, amid encouraging roars and great final tail-wagging.

At Glasgow, in 1876, I was President of the Biological Section, and our meeting was rendered rather lively by the announcement of a paper by Professor W. F. Barrett on experiments in thought-reading. The reading of this was opposed by Dr. W. B. Carpenter and others, but as it had been accepted by the section, it was read. Then followed a rather heated discussion; but there were several supporters of the paper, among whom was Lord Rayleigh, and the public evidently took the greatest interest in the subject, the hall being crowded. After having studied the matter some years longer, Professor Barrett, with the assistance of the late Frederick Myers, Professor Sidgwick, Edmund Gurney, and a few other friends, founded the Society for

Psychical Research, which has collected a very large amount of evidence and is still actively at work.

My wife and I were entertained at Glasgow by Mr. and Mrs. Mirlees, and at one of their dinner-parties we enjoyed the company of William Pengelly, of Torquay, the well-known explorer of Kent's Cavern, whose acquaintance I had made some years before while spending a few days at Torquay with my friend and publisher, Mr. A. Macmillan. He sat on one side of our hostess, and my wife and I on the other, and during the whole dinner he kept up such a flow of amusing and witty conversation that the entire party (a large one) looked at us with envy. He was certainly among the most genial and witty men I have ever met, and could make even dry scientific subjects attractive by his humorous way of narrating them.

After this year I felt that I had pretty well exhausted the interests of the Association meetings, and preferred to take my autumn holiday, with my wife and two children, either by the sea or among the mountains, where we could quietly enjoy the beauties of nature in aspects somewhat new to us; the only exception I afterwards made being the jubilee meeting at York, and even here the chief attractions were the beautiful Alpine gardens of Mr. Backhouse, the excursion to Rievaulx Abbey, and a visit afterwards to my friend Dr. Spruce in his retirement at Welburn, near Castle Howard.

CHAPTER XVII

HOME LIFE AND WORK

(1870–1885)

IN March, 1870, I took an old cottage at Barking, where I was still almost in London. Though Barking was a miserable village, surrounded by marshes and ugly factories, there were yet some pleasant walks along the Thames and among the meadows, while within a quarter of a mile of us was a well-preserved tumulus close to an old farmhouse. Here, too, we had some very pleasant neighbours. Sir Antonis Brady at Stratford, whom I had often visited with my friend Silk, and who had a fine collection of fossils from the gravels of the district; Mr. C. M. Ingleby, the Shakespearean commentator, who was interested in spiritualism; and more especially Colonel Hope, V.C., who was living at Parsloes, an old manor house within an easy walk, and with whose amiable and intellectual family we spent many pleasant Sunday afternoons.

In the following year I found near the village of Grays,

on the Thames, twenty miles from London, a picturesque old chalk-pit which had been disused so long that a number of large elms and a few other trees had grown up in its less precipitous portions. The chalk here was capped by about twenty feet of Thanet sand and pleistocene gravel, and from the fields at the top there was a beautiful view over Erith to the Kent hills and down a reach of the Thames to Gravesend, forming a most attractive site for a house After some difficulty I obtained a lease for ninety-nine years of four acres, comprising the pit itself, an acre of the field on the plateau above, and about an equal amount of undulating cultivable ground between the pit and the lane which gave access to it. I had to pay seven pounds an acre rent, as the owner could not sell it, and though I thought it very dear, as so much of it was unproductive, the site was so picturesque, and had such capabilities of improvement, that I thought it would be a fair investment.

THE DELL, GRAYS.
(*Built for A. R. Wallace.*)

As there was a deep bed of rough gravel on my ground and there were large cement works at Grays, I thought it would be economical to build of concrete, and I found an architect of experience, who made the plans and specifications, while I myself saw that the gravel was properly washed. In order to obtain water in ample quantity for building and also for garden and other purposes, I had a well sunk about a hundred feet into a water-bearing stratum of the chalk, and purchased a small iron windmill with a two-inch force pump to obtain the water. I made two small concrete ponds in the garden—one close to the windmill—and a large tank at the top of a low tower to supply house water. My friend Geach, the mining engineer whom I had met in Timor and Singapore, was now at home, and took an immense interest in my work. He helped me to find the windmill—the only one that we could discover in any of the engineering shops in London—and the well being completed, he and I, with the assistance of my gardener, did all the work of fitting the pump at the bottom of the well with connecting-rods and guides up to the windmill, which also we erected and set to work ourselves. As the windmill had no regulating apparatus, and, when the wind became strong, revolved far too rapidly, and even bent the connecting-rod, I attached to the ends of the iron vanes pieces of plate iron about a foot square, fixed at right angles to the line of motion. These acted as brakes as soon as the revolution became moderately rapid, but had little effect when it was slow; and the arrangement worked very well.

We got into the house in March, 1872, and I began to take that pleasure in gardening, and especially in growing uncommon and interesting as well as beautiful plants, which in various places, under many difficulties and with mingled failures and successes, has been a delight and solace to me ever since.

At this time I was somewhat doubtful in what particular direction to work, as I found that I could not now feel sufficient interest in any branch of systematic zoology to devote myself to the minute study required for the classification and description of any important portion of my collections. There were many other men who could do that better than I could, while my special tastes led me to some work which involved a good deal of reasoning and generalization. It was, I think, my two friends, Professor A. Newton and Dr. Sclater, who urged me to undertake a general review of the geographical distribution of animals, and after a little discussion of the subject I came to the conclusion that I might perhaps be able to do it; although, if I had been aware of the difficulties of the task, I should probably not have undertaken it.

As this was the largest and perhaps the most important scientific work I have done, I may perhaps be allowed here to say a few words as to its design and execution. I had already, in several of my papers and articles, explained my general views of the purport and scope of geographical distribution as a distinct branch of biological science. I had accepted and supported Dr. P. L. Sclater's division of the earth's surface into six great zoological regions, founded upon a detailed examination of the distribution of birds, but equally applicable to mammalia, reptiles, and several other great divisions, and best serving to illustrate and explain the diversities and apparent contradictions in the distribution of all land animals; and I may now add that the additional facts accumulated, and the various divisions suggested during the thirty years that have since elapsed, have not in the least altered my opinions on this matter.

In whatever work I have done I have always aimed at systematic arrangement and uniformity of treatment throughout. But here the immense extent of the subject, the overwhelming mass of detail, and above all the excessive diversities in the amount of knowledge of the different classes of animals, rendered it quite impossible to treat all alike. My preliminary studies had already satisfied me that it was quite useless to attempt to found any conclusions on those groups which were comparatively little known, either as regards the proportion of species collected and described, or as regards their systematic classification. It was also clear that as the present distribution of animals is necessarily due to their past distribution, the greatest importance must be given to those groups whose fossil remains in the more recent strata are the most abundant and the best known. These considerations led me to limit my work in its detailed systematic groundwork, and study of the principles and laws of distribution, to the mammalia and birds, and to apply the principles thus arrived at to an explanation of the distribution of other groups, such as reptiles, fresh-water fishes, land and fresh-water shells, and the best-known insect orders. The work, entitled "The Geographical Distribution of Animals," was published in 1876, in two thick volumes, and it occupied a good deal of my time during the four years I lived at Grays.

No one is more aware than myself of the defects of the work, a considerable portion of which are due to the fact that it was written a quarter of a century too soon—at a time when both zoological and palæontological discovery were advancing with great rapidity, while new and improved classifications of some of the great classes and orders were in constant progress. But though many of the details given in these volumes would now require alteration, there is no reason to believe that the great features of the work and general principles established by it will require any important modification. Its most severe critics are our American cousins, who, possessing a "region" of their own, have been able to explore it very rapidly; while from several references made to it, I think it is appreciated on the European continent more than it is in our own country.

A. R. WALLACE. AGED 55.

While this work was in progress I wrote a considerable number of reviews and articles, published my book on "Miracles and Modern Spiritualism," and wrote the articles "Acclimatization" for the "Encyclopædia Britannica."

In 1876 I sold the house at Grays and removed to Dorking, where we lived for two years. But finding the climate relaxing, we moved next to Croydon, chiefly in order to send our children first to a kindergarten, and then to a high school, and remained there till May, 1881.

During this period, besides my usual reviews and articles, I prepared my address as president of the Biological Section of the British Association at Glasgow, wrote the article on "Distribution— Zoology" for the "Encyclopædia Britannica," and prepared a volume on "Tropical Nature," which was published in 1878. In this work I gave a general sketch of the climate, vegetation, and animal life of the equatorial zone of the tropics from my own observations in both hemispheres.

As soon as we were settled at Croydon, I began to work at a volume which had been suggested to me by the necessary limitations of my "Geographical Distribution of Animals." In that work I had, in the first place, dealt with the larger groups, coming down to families and genera, but taking no account of the various problems raised by the distribution of particular *species*. In the next place, I had taken little account of the various islands of the globe, except as forming subregions or parts of subregions. But I had long seen the great interest and importance of these, and especially of Darwin's great discovery of the two classes into which they are naturally divided—oceanic and continental islands. I had already given lectures on this subject, and had become aware of the great interest attaching to them, and the great light they threw upon the means of dispersal of animals and plants, as well as upon the past changes, both physical and biological, of the earth's surface. In the third place, the means of dispersal and colonization of animals is so connected with, and often dependent on, that of plants, that a consideration of the latter is essential to any broad views as to the distribution of life upon the earth, while they throw unexpected light upon these exceptional means of dispersal which, because they are exceptional, are often of paramount importance in leading to the production of new species and in thus determining the nature of insular floras and faunas.

Having no knowledge of scientific botany, it needed some courage, or, as some may think, presumption, to deal with this aspect of the problem; but, on the other hand, I had long been excessively fond of plants, and was always interested in their distribution. The subject, too, was easier to deal with, on account of the much more complete knowledge of the detailed distribution of plants than of animals, and also because their classification was in a more advanced and stable condition. Again, some of the most interesting of the islands of the globe had been carefully studied botanically by such eminent botanists as Sir Joseph Hooker, for the Galapagos, New Zealand, Tasmania, and the Antarctic islands; Mr. H. C. Watson for the Azores; Mr. J. G. Baker for Mauritius and other Mascarene islands; while there were floras by competent botanists of the Sandwich Islands, Bermuda, and St. Helena. With such excellent material, and with the further assistance of Sir Joseph Hooker's invaluable essays on the relations of the southern and northern floras, I felt that my work would be mainly of a statistical nature, as interpreted by those general principles of organic evolution which were my especial study.

But I also found it necessary to deal with a totally distinct branch of science—recent changes of climate as dependent on changes of the earth's surface, including the causes and effects of the glacial epoch, since these were among the most powerful agents in causing the dispersal of all kinds of organisms, and thus bringing about the actual distribution that now prevails. This led me to a careful study of Mr. James Croll's remarkable works on the subject of the astronomical causes of glacial and interglacial periods, and I had much correspondence with him on difficult points of his theory. While differing on certain details, I adopted the main features of his theory, combining with it the effects of changes in height and extent of land which form an important adjunct to the meteorological agents. To this subject I devoted two of my longest and most argumentative chapters, introducing many considerations not before taken account of, and leading, I still think, to a more satisfactory explanation of the causes that actually brought about the glacial epoch than any which have since been put forth.

Besides this partially new theory of the causes of glacial epochs, the work contained a fuller statement of the various kinds of evidence proving that the great oceanic basins are permanent features of the earth's surface, than had before been given; also a discussion of the mode of estimating the duration of geological periods, and some considerations leading to the conclusion that organic change is now less rapid than the average, and therefore that less time is required for this change than has hitherto been thought necessary. I was also, I believe, the first to point out the great differences between the more ancient continental islands and those of more recent origin, with the interesting conclusions as to geographical changes afforded by both; while the most important novelty is the theory by which I explained the occurrence of northern groups of plants in all parts of the southern hemisphere—a phenomenon which Sir Joseph Hooker had pointed out, but had then no means of explaining.

This volume, on "Island Life," involved much detailed work as regards the species of plants and animals, information

347

on which points I had to obtain from numerous specialists, involving a great amount of correspondence; while it was illustrated by a large number of maps and diagrams, most of which were drawn by myself. The preparation and writing this book occupied me for about three years, and it was published in 1880. It has gone through three editions, which have involved a large amount of corrections and additions; and it is a work which seems to have opened up a new world of interesting fact and theory to a large number of readers, from several of whom I have received letters expressing the delight and instruction it has given them.

In 1878 I wrote a volume on Australasia for Stanford's "Compendium of Geography and Travel," in which I gave a fuller account than usual of the physical geography, the natural history, and the geology of Australia. In a later edition of this work, in 1893, I gave a much fuller account of the natives of Australia, and adduced evidence for the theory that they are really a primitive type of the great Caucasian family of mankind, and are by no means so low in intellect as has been usually believed. This view seems now to be generally accepted.

In 1881 a society was formed for advocating the nationalization of the land, of which I was elected president, and in 1882 I published a volume, entitled "Land Nationalization: its Necessity and its Aims." Some account of this movement will be given in a

NUTWOOD COTTAGE, GODALMING

future chapter. Its publication brought me letters of sympathy and general agreement from Sir David Wedderburn, M.P., Lord Mount-Temple, and many other friends and correspondents.

In this year, on June 29, the Dublin University gave me the honorary degree of LL.D. I will here give the very short but flattering Latin speech of the public orator in introducing me, with a translation by my friend Mr. Comerford Casey—

"Introduco quoque ALFREDUM RUSSEL WALLACE, Darwinii aemulum, immo Darwinium alterum. Neque hunc neque ilium variae eluserunt species atque ora ferarum. Darwinius nempe lauri foetus auricomos decerpsit primus. Sed quid querimur?

'Primo avulso non deficit alter Aureus, et simili frondescit virga metallo.'"

I introduce also Alfred Russel Wallace, the friendly rival of Darwin, indeed, a second Darwin. Equally familiar to both are the different species and varieties of animals. Darwin, indeed was the first to pluck the golden laurel-branch. Yet through this did Wallace suffer no eclipse; for as Virgil sang—

'One branch removed, another was to hand: Another, bright and golden as the first.'"

In the year 1881 I removed to Godalming, where I had built a small cottage near the water-tower and at about the same level as the Charterhouse School. We had been partly induced to come here to be near my very old friend Mr. Charles Hayward, whom I had first known during my residence at Neath about forty years before. He was living with his nephew, the late C. F. Hayward, a well-known architect, whose children were about the same age as my own. We found here some very pleasant friends among the masters at Charterhouse School, as well as among residents who had come to the place for its general educational advantages or for the charm of its rural scenery. We had here about half an acre of ground with oak trees and hazel bushes (from which I named our place "Nutwood Cottage"), and during the eight years we lived there I thoroughly enjoyed making a new garden, in which, and a small greenhouse, I cultivated at one time or other more than a thousand species of plants. The soil was a deep bed of the Lower Greensand formation, with a thin surface layer of leaf-mould, and it was very favourable to many kinds of bulbous plants as well as half-hardy shrubs, several of which grew there more freely and flowered better than in any of my other gardens.

In 1884 Messrs. Pears offered a prize of £100 for the best essay on "The Depression of Trade," and Professor Leone Levi had agreed to be one of the judges. As I had been for some time disgusted with the utter nonsense of many of the articles on the subject in the press, while what seemed to me the essential and fundamental causes were never so much as referred to, I determined to compete, though without any expectation of success. The essay was sent in some time during the summer of 1885, and in July I received a letter from Professor Leone Levi, in which he writes: "My colleague and myself were greatly pleased with the essay bearing a motto from Goldsmith. We, however, did not see our way to recommend it for the prize, especially on account of disagreement as to the remedies suggested. But, the essay having great merit, we thought it proper to open the envelope in order to correspond with the author."

He then asked me if I would allow the first part of my essay, upon "Conditions and Causes," to be printed with the other essays.

As my proposed remedies were the logical conclusion from the "Conditions and Causes," which I had detailed, and of which the validity seemed to be admitted, I of course declined this offer, and Messrs. Macmillan agreed to publish it under the title, "Bad Times: An Essay on the Present Depression of Trade, tracing it to its Sources in enormous Foreign Loans, excessive War Expenditure, the Increase of Speculation and of Millionaires, and the Depopulation of the Rural Districts; with Suggested Remedies."

This little book was widely noticed, but most of the reviewers adverted to the fact that I was an advocate of land-nationalization, and therefore that my proposed remedies were unsound. But a few were more open-minded, and the *Newcastle Chronicle* declared it to be "the weightiest contribution to the subject made in recent times."

CHAPTER XVIII

A LECTURE TOUR IN AMERICA

(1886–1887)

TOWARDS the close of the year 1885 I received an invitation from the Lowell Institute of Boston, U.S.A., to deliver a course of lectures in the autumn and winter of 1886. After some consideration I accepted this, and began their preparation, taking for my subject those portions of the theory of evolution with which I was most familiar. At this time I had made the acquaintance of the Rev. J. G. Wood, the well-known writer of many popular works on natural history. He had been twice on lecturing tours to America, and gave me some useful information, besides recommending an agent he had employed and who had arranged lectures for him at various schools and colleges. I had already lectured in many English towns on the permanence of the great oceans, on oceanic and continental islands, and on various problems of geographical distribution. To these subjects I now added one on "The Darwinian Theory," illustrated by a set of original diagrams of variation. I also wrote three lectures on the "Colours of Animals (and Plants)," dwelling especially on protective colours, warning colours, and mimicry, and for these I had to obtain a series of lantern slides coloured from nature, so as to exhibit the most striking examples of these curious and beautiful phenomena.

I left London on October 9 in a rather slow steamer, in order to have a cabin to myself at a moderate price, and landed at New York on the 23rd, after a cold and disagreeable passage.

My tour in America lasted about ten months, and extended from New York, Boston, and Washington to San Francisco, and back to Montreal and Quebec. I visited and lectured at many of the great cities, and met many well-known and intellectual people, including amongst others Oliver Wendell Holmes, James Russell Lowell, and Edward Waldo Emerson, son of the philosopher. I also had the opportunity of seeing some of the finest scenery and most interesting objects in North America.

Whilst in California I once more had the pleasure of meeting my brother John and his wife, neither of whom I had seen for nearly forty years.

As American cities have been so frequently described, I will here only give an account of a few of the interesting sights or natural objects which most impressed me, with some remarks on my general impression of the country and people.

The first excursion I made in America was a trip up the Hudson River to West Point, passing the celebrated "Palisades"—a continuous row of cliffs about two hundred feet high, and extending for nearly twenty miles on the south bank of the river. They look exactly like a huge fence of enormous split trees, placed vertically, side by side, but are really basaltic columns like those at the Giant's Causeway, crowning a slope of fallen rock. In places the well-wooded country was very beautiful, with the autumnal tints of bright red, purple, and yellow, though we were a little late to see them in perfection. Where we landed, I was delighted to see wild vines clambering over the trees, as well as the Virginia creeper, and there were also sumachs and other characteristic American plants. The situation of the great American Military College is splendid, on an elevated promontory in a bend of the Hudson, surrounded by rugged wooded hills, and with magnificent views up and down the river.

In March, 1887, on my way back from Toronto (where I had been to lecture) to Washington, I spent four days at Niagara, living at the old hotel on the Canadian side, in a room that looked out on the great fall, and where its continuous musical roar soothed me to sleep. It was a hard frost, and the American falls had great ice-mounds below them, and ranges of gigantic icicles near the margins. At night the sound was like that of a strong, steady wind at sea, but even more like the roar of the London streets heard from the middle of Hyde Park. When in bed a constant vibration was felt. I spent my whole time wandering about the falls, above and below, on the Canadian and the American sides, roaming over Goat Island and the Three Sisters Islands far in the rapids above the Horse-shoe Fall, which are almost as impressive as the fall itself. The small Luna Island dividing the American falls was a lovely sight; the arbor-vitæ trees (*Thuya Americana*), with which it is covered, young and old, some torn and jagged, but all to the smallest twigs coated with glistening ice from the frozen spray, looked like groves of gigantic tree corals, forming the most magnificent and fairylike scene I have ever beheld. All the islands are rocky and picturesque, the trees draped with wild vines and Virginia creepers, and afford a sample of the original American forest vegetation of very great interest. During these four days I was almost entirely alone, and was glad to be so. I was never tired of the ever-changing aspects of this grand illustration of natural forces engaged in modelling the earth's surface. Usually the centre of the great falls, where the depth and force of the water are greatest, is hidden by the great column of spray which rises to the height of four hundred or five hundred feet; but occasionally the wind drifts it aside, and allows the great central gulf of falling water to be seen nearly from top to bottom—a most impressive sight.

Before I left Washington, Judge Holman, with whose

family I sat at meals, took me one morning to call upon the President, Mr. Cleveland. The judge told him I was going to visit California, and that turned the conversation on wine, raisins, etc., which did not at all interest me. There was no ceremony whatever, but, of course, I had nothing special to say to him, and he had nothing special to say to me, the result being that we were both rather bored, and glad to get it over as soon as we could. I then went to see the White House, some of the reception-rooms of which were very fine; but there was a great absence of works of art, the only paintings I saw being portraits of Washington and his wife.

Washington itself is a fine and even picturesque city, owing to its designer having departed from the rigid rectangularity of most American cities by the addition of a number of broad diagonal avenues crossing the rectangles at different angles, and varying from one to four miles long. The broadest of these are one hundred and sixty feet wide, planted with two double avenues of trees, and with wide grassy spaces between the houses and the pavements. Wherever these diagonal avenues intersect the principal streets, there are quadrangular open spaces forming gardens or small parks, planted with shrubs and trees, and with numerous seats.

When I left Washington for Cincinnati I broke my journey to visit the caves at Louray, which had been discovered about ten years before. I and a few other visitors walked through the best parts (which are lit up with electric lights) for about two hours, through a variety of passages, galleries, and halls, some reaching a hundred feet in height, some having streams or pools of water, and some chasms of unknown depth, such as most limestone caves possess. Everywhere there are stalactites of the most varied forms, and often of the most wonderful beauty. Usually they form pillars as of some strange architecture, sometimes they hang down like gigantic icicles, one of these over sixty feet long being with its dripping apex only a few inches from the floor. In some places the stalactites resemble cascades, in others organs, and several are like statues, and have received appropriate names. Many of them are most curiously ribbed; others, again, have branches growing out of them at right angles a few inches long—a most puzzling phenomenon. There is a Moorish tent, in which fine white drapery hangs in front of a cave, a ballroom beautifully ornamented with snow-white stalactitic curtains, etc. Some of these, when struck, give out musical notes, and a tune can be played on them. A photograph of the Moorish tent and the curious pillars near it is here reproduced. The curtain is like alabaster, and when a lamp is held behind it, the effect is most beautiful. In many places there are stalagmitic floors, beneath which is clay filled with bones of bats, etc., and at one spot human bones are

THE MOORISH TENT, LOURAY CAVERN.

embedded in the floor under a chasm opening above. The print of an Indian mocassin is also shown petrified by the stalagmite. Rats and mice are found with very large eyes; and there are some blind insects and centipedes, as in the Mammoth Cave. Several miles of caverns and passages have already been explored, but other wonders may still be hidden in its deeper recesses.

From Cincinnati I continued my journey westward, paying short visits to Kansas City, Denver, and Salt Lake City, and then on to San Francisco, where I met my brother John.

A few days after my arrival in San Francisco I was taken for a drive by a friend, Dr. Gibbons, into the foothills to see the remains of the Redwood forest that once covered them, but which had all been ruthlessly destroyed to supply timber for the city and towns around. We wound about among the hills and valleys, all perfectly dry, till we reached a height of fifteen hundred feet, where many clumps of young redwoods were seen, and, stopping at one of these, Dr. Gibbons took me inside a circle of young trees from twenty to thirty feet high, and showed me that they all grew on the outer edge of the huge charred trunk of an old tree that had been burnt down. This stump was thirty-four feet in diameter, or quite as large as the very largest of the more celebrated Big Trees, the *Sequoia gigantea*. The doctor had searched all over these hills, and this was the largest stump he had found, though there were numbers between twenty and thirty feet. The tree derives its botanical name, *sempervirens* (ever-growing) from the peculiar habit of producing young trees from the burnt or decayed roots of the old trees. These enormous trees, being too large to cut down, were burnt till sufficiently weakened to fall, and this particular tree had been so burnt about forty years before.

While staying with my brother in Stockton, Cal., I went with him and his daughter to spend a few days in the Yosemite Valley. The journey there— two hours by rail and two days by coach—was very interesting, but often terribly dusty. The first day we were driving for nine hours in the foothills, among old mining camps with their ruined sheds and reservoirs and great gravel-heaps, now being gradually overgrown by young pines and shrubs. Here and there we passed through bits of forest with tall pines and shrubby undergrowth, but generally the country was bare of fine trees, scraggy, but burnt up, and the roads insufferably dusty. At 9 p.m. we reached Priest's (two thousand five hundred feet elevation), where we had supper, bed, and breakfast.

Next day was much more enjoyable. The road was wonderfully varied, always going up or down, diving into deep wooded valleys with clear and rapid streams, then up the slope, winding round spurs, crossing ridges, and down again into valleys, but always mounting higher and higher. And as we got deeper into the sierra, the vegetation continually changed, the pines became finer both in form, size, and beauty. At about three thousand feet we first saw the beautiful Douglas fir, and the cedar (*Libocedrus decurrens*), both common in our gardens; then still higher there were silver firs and the fine *Picea nobilis,* as well as a few of the Big Trees (*Sequoia gigantea*), the road being cut right through the middle of one of these (at about five thousand eight hundred feet). From the summit we descended towards the valley, and then down a steep zigzag road, with the beautiful

Bridal Veil Fall opposite, and the grand precipice of El Capitan before us, then into the valley itself with its rushing river, to the hotel in the dusk.

As both hotel and excursions were here very costly, we only stayed two clear days, and went one "excursion" to the Nevada Fall, the grandest, if not the most beautiful, in the valley. My brother and niece rode up, but I walked to enjoy the scenery, and especially the flowers and ferns and the fine glaciated rocks of the higher valley. The rest of my time I spent roaming about the valley itself and some of its lower precipices, looking after its flowers, and pondering over its strange, wild, majestic beauty and the mode of its formation. On the latter point I have given my views in an article on "Inaccessible Valleys," reprinted in my "Studies." The hotel dining-room looks out upon the Yosemite Falls, which, seen one behind the other, have the appearance of a single broken cascade of more than two thousand five hundred feet. I walked up about a thousand feet to get a nearer view of the upper fall, which, in its ever-changing vapour-streams and water-rockets, is wonderfully beautiful. To enjoy this valley and its surroundings in perfection, a small party should come with baggage-mules and tents, as early in the season as possible, when the falls are at their grandest and the flowers in their spring beauty, and when, by camping at different stations in the valley and in the mountains and valleys around it, all its wonderful scenes of grandeur and beauty could be explored and enjoyed. It is one of the regrets of my American tour that I was unable to do this.

On our way back I turned off at the foot of the hills to visit the Calaveras Grove of big trees which my brother and niece had seen before, and I had to sleep on the way. I stayed three days, examining and measuring the trees, collecting flowers, and walking one day to the much larger south grove six miles off, where there are said to be over a thousand full-grown trees. The walk was very interesting, over hill and valley, through forest all the way, except one small clearing. At a small rocky stream I found the large *Saxifraga peltata* growing in crevices of rocks just under water, and I passed numbers of fine examples of all the chief pines, firs, and cypresses. At the grove there were numbers of very fine trees, but none quite so large as the largest in the Calaveras Grove. Many of them have names. "Agassiz" is thirty-three feet wide at base, and has an enormous hole burnt in it eighteen feet wide and the same depth, and extending upwards ninety feet like a large cavern; yet the tree is in vigorous growth. The Sequoias are here thickly scattered among other pines and firs, sometimes singly, sometimes in groups of five or six together. There are many twin trees growing as a single stem up to twenty or thirty feet, and then dividing. But the chief feature of this grove is the abundance of trees to be seen in every direction, of large or moderate size, and with clean, straight stems showing the brilliant orange-brown tint and silky or plush-like glossy surface, characteristic of the bark of this noble tree when in full health and vigorous growth. In no forest that I am acquainted with is there any tree with so beautiful a bark or with one so thick and elastic.

In the chapter on "Flowers and Forests of the Far West" (in my "Studies"), I have given a summary of the chief facts known about these trees, with particulars of their dimensions and probable age. I need not, therefore, repeat these particulars here.

But of all the natural wonders I saw in America, nothing impressed me so much as these glorious trees. As with Niagara, their majesty grows upon one by living among them. The forests of which they form a part contain a number of the finest conifers in the world—trees that in Europe or in any other Northern forest would take the very first rank. These grand pines are often from two hundred to two hundred and fifty feet high, and seven or eight feet in diameter at five feet above the ground, where they spread out to about ten feet. Looked at alone, these, are noble trees, and there is every gradation of size up to these. But the Sequoias take a sudden leap, the average full-grown trees being twice this diameter, and the largest three times the diameter of these largest pines; so that when they were first found the accounts of the discoveries were disbelieved. My brother told me an interesting story of this discovery. The early miners used to keep a hunter in each camp to procure game for them, venison, and especially bear's meat, being highly esteemed. These men used to search the forests for ten or twenty miles round the camps while hunting. The hunter of the highest camp on the Stanislaus river came home one evening, and after supper told them of a big tree he had found that beat all he had ever seen before. It had three times as big a trunk as any tree within ten miles round. Of course they all laughed at him, told him they were not fools: they knew what trees were as well as he did; and so on. Then he offered to *show* it them, but none would go; they would not tramp ten or twelve miles to be made fools of. So the hunter had to bide his time. A week or two afterwards he came home one Saturday night with a small bag of game; but he excused himself by saying that he had got the finest and fattest bear he had ever killed, and as next day was Sunday he thought that six or eight of them would come with him and bring the meat home.

The next morning a large party started early, and after a long walk the hunter brought them suddenly up to the big tree, and, clapping his hand on it, said, "Here's my fat bear. When I called it a tree, you wouldn't believe me. Who's the fool now?" This was the great pavilion tree of the Calaveras Grove, twenty-six feet in diameter at five feet from the ground—over eighty feet in circumference, so that it would require fourteen tall men with arms outstretched to go round it. This tree was afterwards cut down by boring into the trunk at six feet from the ground with long pump-augers from each side, so as to meet in the centre. The first fourteen feet was then cut into sections, and one supplied to each of the older States. The rest remains as it fell, and can be walked on to a distance of about two hundred and ten feet from the stump, and here it is still six feet in diameter. To examine this wonderful wreck of the grandest tree then living on our globe is most impressive. The rings on the stump of this tree have been very carefully counted by Professor Bradley, of the University of California, and were found to be 1240, which no doubt gives the age of the tree very accurately, as the winters are here severe, and the season of growth very well marked.

On my return journey I crossed the Rocky Mountains by the Denver and Rio Grande Railway, passing, through some of the wildest and most magnificent country. In places there were precipices about a thousand feet high, side cañons like narrow slits or winding majestic ravines, often with vertical walls, and the river roaring and raging in a tumultuous flood close alongside of us.

I broke my journey at Denver, where I made an excursion to Gray's Peak, one of the highest in the range. I also paid a visit to the celebrated "Garden of the Gods" with its fantastic weather-worn and brilliantly coloured rocks.

I next went on to Chicago, where I spent only a few hours while waiting for my train to Montreal and Quebec whence I set sail for England.

Before closing this brief account of my tour in America, it may be interesting to record my general impressions as to the country and people. In my journal I find this note: "During more than ten months in America, taking every opportunity of exploring woods and forests, plains and mountains, deserts and gardens, between the Atlantic and Pacific coasts, and extending over ten degrees of latitude, I never once saw either a humming-bird or a rattlesnake, or even any living snake of any kind. In many places I was told that humming-birds were usually common in their gardens, but they hadn't seen any this year! This was my luck. And as to the rattlesnakes, I was always on the look out in likely places, and there are plenty still, but they are local. I was told of a considerable tract of land not far from Niagara which is so infested with them that it is absolutely useless. The reason is that it is very rocky, with so many large masses lying about overgrown with shrubs and briars as to afford them unlimited hiding-places, and the labour of thoroughly clearing it would be more costly than the land would be worth."

The general impression left upon my mind as to the country itself is the almost total absence of that simple rural beauty which has resulted, in our own country and in some other parts of Europe, from the very gradual occupation of the land as it was required to supply food for the inhabitants, together with our mild winters allowing of continuous cultivation, and the use in building of local materials adapted to the purposes required by handwork, instead of those fashioned by machinery. This slow development of agriculture and of settlement has produced almost every feature which renders our country picturesque or beautiful: the narrow winding lanes, following the contours of the ground; the ever-varying size of the enclosures, and their naturally curved boundaries; the ditch and bank and the surmounting hedgerow, with its rows of elm, ash or oak, giving variety and sylvan beauty to the surroundings of almost every village or hamlet, most of which go back to Saxon times; the farms or cottages built of brick, or stone, or clay, or of rude but strong oak framework filled in with clay or lath and roughcast, and with thatched or tiled roofs, varying according to the natural conditions, and in all showing the slight curves and irregularities due to the materials used and the hand of the worker;—the whole, worn and coloured by age and surrounded by nature's grandest adornment of self-sown trees in hedgerow or pasture, combine together to produce that charming and indescribable effect we term picturesque. And when we add to these the numerous footpaths which enable us to escape the dust of high-roads and to enjoy the glory of wild flowers which the innumerable hedgerows and moist ditches have preserved for us, the breezy downs, the gorse-clad commons and the heath-clad moors still unenclosed, we are, in some favoured districts at least, still able thoroughly to enjoy all the varied aspects of beauty which our country affords us, but which are, alas! under the combined influences of capitalism and landlordism, fast disappearing.

But in America, except in a few parts of the northeastern States, none of these favourable conditions have prevailed. Over by far the greater part of the country there has been no natural development of lanes and tracks and roads as they were needed for communication between villages and towns that had grown up in places best adapted for early settlement; but the whole country has been marked out into sections and quarter-sections (of a mile, and a quarter of a mile square), with a right of way of a certain width along each section-line to give access to every quarter-section of one hundred and sixty acres, to one of which, under the homestead law, every citizen had, or was supposed to have, a right of cultivation and possession. Hence, in all the newer

States there are no roads or paths whatever beyond the limits of the townships, and the only lines of communication for foot or horsemen or vehicles of any kind are along these rectangular section-lines, often going up and down hill, over bog or stream, and almost always compelling the traveller to go a much greater distance than the form of the surface rendered necessary.

Then again, owing to the necessity for rapidly and securely fencing in these quarter-sections, and to the fact that the greater part of the States first settled were largely forest-clad, it became the custom to build rough, strong fences of split-trees, which utilized the timber as it was cut and involved no expenditure of cash by the settler. To avoid the labour of putting posts in the ground the fence was at first usually built of rails or logs laid zigzag on each other to the height required, so as to be self-supporting, the upper pairs only being fastened together by a spike through them, the waste of material in such a fence being compensated by the reduction of the labour, since the timber itself was often looked upon as a nuisance to be got rid of before cultivation was possible. This fact of timber being in the way of cultivation and of no use till cut down, led to the very general clearing away of all the trees from about the house, so that it is a comparatively rare thing, except in the eastern towns and villages, to find any old trees that have been left standing for shade or for beauty.

For these and for similar causes acting through the greater part of North America, there results a monotonous and unnatural ruggedness, a want of harmony between man and nature, the absence of all those softening effects of human labour and human occupation carried on for generation after generation in the same simple way, and in its slow and gradual utilization of natural forces allowing the renovating agency of vegetable and animal life to conceal all harshness of colour or form, and clothe the whole landscape in a garment of perennial beauty.

Over the larger part of America everything is raw and bare and ugly, with the same kind of ugliness with which we also are defacing our land and destroying its rural beauty. The ugliness of new rows of cottages built to let to the poor, the ugliness of the mean streets of our towns, the ugliness of our "black countries "and our polluted streams. Both countries are creating ugliness, both are destroying beauty; but in America it is done on a larger scale and with a more hideous monotony. The more refined among the Americans see this themselves as clearly as we see it. One of them has said, "A whole huge continent has been so touched by human hands that, over a large part of its surface it has been reduced to a state of unkempt, sordid ugliness; and it can be brought back into a state of beauty only by further touches of the same hands more intelligently applied."[1]

Turning now from the land to the people, what can

we say of our American cousins as a race and as a nation? The great thing to keep in mind is, that they are, largely and primarily, of the same blood and of the same nature as ourselves, with characters and habits formed in part by the evil traditions inherited from us, in part by the influence of the new environment to which they have been exposed. Just as we owe our good and bad qualities to the intermixture and struggle of somewhat dissimilar peoples, so do they. Briton and Roman, Saxon and Dane, Norsemen and Norman-French, Scotch and Irish Celts—all have intermingled in various proportions, and helped to create that energetic amalgam known the world over as Englishmen. So North America has been largely settled by the English, partly by Dutch, French, and Spanish, whose territories were soon absorbed by conquest or purchase; while, during the last century, a continuous stream of immigrants—Germans, Irish, Highland and Lowland Scotch, Scandinavians, Italians, Russians—has flowed in, and is slowly but surely becoming amalgamated into one great Anglo-American people.

[1] *The Century, June, 1887.*

Most of the evil influences under which the United States have grown to their present condition of leaders in civilization, and a great power among the nations of the world, they received from us. We gave them the example of religious intolerance and priestly rule, which they have now happily thrown off more completely than we have done. We gave them slavery, both white and black—a curse from the effects of which they still suffer, and out of which a wholly satisfactory escape seems as remote as ever. But even more insidious and more widespread in its evil results than both of these, we gave them our bad and iniquitous feudal land system; first by enormous grants from the Crown to individuals or to companies, but also—what has produced even worse effects—the ingrained belief that *land*—the first essential of life, the source of all things necessary or useful to mankind, by labour upon which all wealth arises—may yet, justly and equitably, be owned by individuals, be monopolized by capitalists or by companies, leaving the great bulk of the people as absolutely dependent on these monopolists for permission to work and to live as ever were the negro slaves of the South before emancipation.

The result of acting upon this false conception is, that the Government has already parted with the whole of the accessible and cultivable land, and though large areas still remain for any citizen who will settle upon it, by the mere payment of very moderate fees, this privilege is absolutely worthless to those who most want it—the very poor. And throughout the western half of the Union one sees everywhere the strange anomaly of building lots in small remote towns, surrounded by thousands of uncultivated acres (and perhaps ten years before sold for eight or ten shillings an acre), now selling at the rate of from £1000 to £20,000 an acre ! It is not an uncommon thing for town lots in new places to double their value in a month, while a fourfold increase in a year is quite common. Hence land speculation has become a vast organized business over all the Western States, and is considered to be a proper and natural mode of getting rich. It is what the Stock Exchange is to the great cities. And this wealth, thus gained by individuals, initiates that process which culminates in railroad and mining kings, in oil and beef trusts, and in the thousand millionaires and multi-millionaires whose vast accumulated incomes are, every penny of them, paid by the toiling workers, including the five million of farmers whose lives of constan toil only result for the most part in a bare livelihood, while the railroad magnates and corn speculators absorb the larger portion of the produce of their labour.

What a terrible object-lesson is this as to the fundamental wrong in modern societies which leads to such a result! Here is a country more than twenty-five times the area of the British Islands, with a vast extent of fertile soil, grand navigable waterways, enormous forests, a superabounding wealth of minerals—everything necessary for the support of a population twenty-five times that of ours—about fifteen hundred millions—which has yet, in little more than a century, destroyed nearly all its forests, is rapidly exhausting its marvellous stores of natural oil and gas, as well as those of the precious metals; and as the result of all this reckless exploiting of nature's accumulated treasures has brought about overcrowded cities reeking with disease and vice, and a population which, though only one-half greater than our own, exhibits all the pitiable phenomena of women and children working long hours in factories and workshops, garrets and cellars, for a wage which will not give them the essentials of mere healthy animal existence; while about the same proportion of its workers, as with us, endure lives of excessive labour for a bare livelihood, or constitute that crying disgrace of modern civilization—willing men seeking in vain for honest work, and forming a great army of the unemployed.

What a demonstration is this of the utter folly and stupidity of those blind leaders of the blind who impute all the evils of *our* social system, all *our* poverty and starvation, to over-population! Ireland, with half the population of fifty years ago, is still poor to the verge of famine, and is therefore still over-peopled. And for England and Scotland as well, the cry is still "Emigrate! emigrate! We are over-peopled !" But what of America, with twenty-five times as much land as we have, and with even greater natural resources, and with a population even more ingenious, more energetic, and more hard-working than ours? Are they over-populated with only twenty people to the square mile? There is only one rational solution of this terrible problem. The system that allows the land and the minerals, the means of communication, and all other public services, to be monopolized for the aggrandisement of the few—for the creation of millionaires— necessarily leads to the poverty, the degradation, the misery of the many.

There never has been, in the whole history of the human race, a people with such grand opportunities for establishing a society and a nation in which the products of the general labour should be so distributed as to produce general well-being. It wanted but a recognition of the fundamental principle of "equality of opportunity," tacitly implied in the Declaration of Independence. It wanted but such social arrangements as would ensure to every child the best nurture, the best training of all its faculties, and the fullest opportunity for utilizing those faculties for its own happiness and for the common benefit. Not only equality before the law, but equality of opportunity, is the great fundamental principle of social justice. This is the teaching of Herbert Spencer, but he did not carry it out to its logical consequence—the inequity, and therefore the social immorality of wealth-inheritance. To secure equality of opportunity there must be no inequality of initial wealth. To allow one child to be born a millionaire and another a pauper is a crime against humanity, and, for those who believe in a deity, a crime against God.[1]

1 I have discussed this subject in my "Studies," vol. ii. chap, xxviii.

It is the misfortune of the Americans that they had such a vast continent to occupy. Had it ended at the line of the Mississippi, agricultural development might have gone on more slowly and naturally, from east to west, as increase of population required. So again, if they had had another century for development before railways were invented, expansion would necessarily have gone on more slowly, the need for good roads would have shown that the rectangular system of dividing up new lands was a mistake, and some of that charm of rural scenery which we possess would probably have arisen.

But with the conditions that actually existed we can hardly wonder at the result. A nation formed by emigrants from several of the most energetic and intellectual nations of the old world, for the most part driven from their homes by religious persecution or political oppression, including from the very first all ranks and conditions of life—farmers and mechanics, traders and manufacturers, students and teachers, rich and poor—the very circumstances which drove them to emigrate led to a natural selection of the *most* energetic, the *most* independent, in many respects the *best* of their several nations. Such a people, further tried and hardened by two centuries of struggle against the forces of nature and a savage population, and finally by a war of emancipation from the tyranny of the mother country, would almost necessarily develop both the virtues, the prejudices, and even the vices of the parent stock in an exceptionally high degree. Hence, when the march of invention and of science (to which they contributed their share) gave them the steamship and the railroad; when California gave them gold and Nevada silver, with the prospect of wealth to the lucky beyond the dreams of avarice; when the great prairies of the West gave them illimitable acres of marvellously fertile soil;—it is not surprising that these conditions with such a people should have resulted in that mad race for wealth in which they have beaten the record, and have produced a greater number of multi-millionaires than all the rest of the world combined, with the disastrous results already briefly indicated.

But this is only one side of the American character. Everywhere there are indications of a deep love of nature, a devotion to science and to literature fully proportionate to that of the older countries; while in inventiveness and in the applications of science to human needs they have long been in the first rank. But what is more important, there is also rapidly developing among them a full recognition of the failings of our common social system, and a determination to remedy it. As in Germany, in France, and in England, the socialists are becoming a power in America. They already influence public opinion, and will soon influence the legislatures. The glaring fact is now being widely recognized that with them, as with all the old nations of Europe, an increase in wealth and in command over the powers of nature such as the world has never before seen, has *not* added to the true well-being of any part of society. It is also indisputable that, as regards the enormous masses of the labouring and industrial population, it has greatly increased the numbers of those whose lives are "below the margin of poverty," while, as John Stuart Mill declared many years ago, it has not reduced the labour of any human being.

An American (Mr. Bellamy) gave us the books that first opened the eyes of great numbers of educated readers to the practicability, the simplicity, and the beauty of Socialism. It is to America that the world looks to lead the way towards a just and peaceful modification of the social organism, based upon a recognition of the principle of Equality of Opportunity, and by means of the Organization of the Labour of all for the Equal Good of all.

CHAPTER XIX

FRIENDS AND OCCUPATIONS OF MY LATER YEARS (1888–1908)

AFTER my return from America in August, 1887, the remainder of the year was occupied at home in overtaking my correspondence, looking after my garden, and making up for lost time in scientific and literary reading, and in considering what work I should next occupy myself with. Many of my correspondents, as well as persons I met in America, told me that they could not understand Darwin's "Origin of Species," but they did understand my lecture on "Darwinism "; and it therefore occurred to me that a popular exposition of the subject might be useful, not only as enabling the general reader to understand Darwin, but also to serve as an answer to the many articles and books professing to disprove the theory of natural selection. During the whole of the year 1888 I was engaged in writing this book, which, though largely following the lines of Darwin's work, contained a great many new features, and dealt especially with those parts of the subject which had been most generally misunderstood.

The spring of 1889 was occupied in passing it through the press, and it was published in May, while a few corrections were made for a second edition in the following October. During this time, however, I gave several of my American lectures in various parts of the country—at Newcastle and Darlington in the spring of 1888; in the autumn at Altrincham and Darwen; and in 1889 at Newcastle, York, Darlington, and Liverpool.

In the autumn of this year the University of Oxford did me the honour of giving me the honorary degree of D.C.L., which I went to receive in November, when I enjoyed the hospitality of my friend Professor E. B. Poulton.

While residing at Godalming, I made the acquaintance of William Allingham and his wife—the poet and the artist—who then lived at Witley—I think it was about the year 1886 or 1887. Mr. Allingham told me that Tennyson wished to see me, and would be glad if I would come some day and lunch with him. A day was fixed, and I accompanied Mr. Allingham to the beautifully situated house on Black-down, near Haslemere, where the poet lived during the summer. Lord Tennyson did not appear till luncheon was on the table, but in the mean time we had seen Lady Tennyson and her son and daughter-in-law, and been shown round the grounds. After luncheon we four men retired to the study, with its three great windows looking south-east over the grand expanse of the finely wooded Weald. Here Tennyson lit his pipe, and we sat round the fire and soon got on the subject of spiritualism, which was evidently what he had wished to talk to me about. I told him some of my experiences, and replied to some of his difficulties—the usual difficulties of those who, though inclined to believe, have *seen* nothing, and find the phenomena as described so different from what they think they ought to be. He was evidently greatly impressed by the evidence, and wished to see something. I gave him the names of one or two mediums whom I believed to be quite trustworthy, but whether he ever had any sittings with them I did not hear.

Then we talked a little about the tropics and of the scenery of the Eastern islands; and, taking down a volume he read, in his fine, deep, chanting voice, his description of Enoch Arden's island—

"The mountain wooded to the peak, the lawns
And winding glades high up like ways to heaven,
The slender coco's drooping crown of plumes,
The lightning flash of insect or of bird,
The lustre of the long convolvuluses
That coiled around the stately stems, and ran
Ev'n to the limit of the land, the glows
And glories of the broad belt of the world,—
All these he saw; but what he fain had seen
He could not see, the kindly human face,
Nor ever hear a kindly voice, but heard
The myriad shriek of wheeling ocean fowl,

The league-long roller thundering on the beach,
The moving whisper of huge trees that branch'd
And blossom'd to the zenith, or the sweep
Of some precipitous rivulet to the wave,
As down the shore he ranged, or all day long
Sat often in the seaward-gazing gorge
A shipwreck'd sailor waiting for a sail
No sail from day to day, but every day
The sunrise broken into scarlet shafts
Among the palms and ferns and precipices;
The blaze upon the waters to the east;
The blaze upon his island overhead;
The blaze upon the waters to the west;
Then the great stars that globed themselves in heaven,
The hollower-bellowing ocean, and again
The scarlet shafts of sunrise—but no sail."

Then he closed the book and asked me if that description was in any way untrue to nature. I told him that so far as I knew from the islands I had seen on the western borders of the Pacific, it gave a strikingly true general description of the vegetation

and the aspects of nature among those islands, at which he seemed pleased. Of course, it avoids much detail, but the amount of detail it gives is correct, and it is just about as much as a rather superior sailor would observe and remember.

We then bade him good-bye, went downstairs and had tea with the ladies, and walked back to Haslemere station. I was much pleased to have met and had friendly converse with the most thoughtful, refined, broad-minded, and harmonious of our poets of the nineteenth century.

Finding my house at Godalming in an unsatisfactory situation, with a view almost confined to the small garden, the south sun shut off by a house and by several oak trees, while exposed to north and east winds, and wishing for a generally milder climate, I spent some weeks in exploring the country between Godalming and Portsmouth, and then westward to Bournemouth and Poole. We were directed by some friends to Parkstone as a very pretty and sheltered place, and here we found a small house to be let, which suited us tolerably well, with the option of purchase at a moderate price. The place attracted us because we saw abundance of great bushes of the evergreen purple veronicas, which must have been a dozen or twenty years old, and also large specimens of eucalyptus; while we were told that there had been no skating there for twenty years. We accordingly took the house, and purchased it in the following year; and by adding later a new kitchen and bedroom, and enlarging the drawing-room, converted it from a cramped, though very pretty cottage, into a convenient, though still small house. The garden on the south side was in a hollow on the level of the basement, while on the north it was from ten to thirty feet higher, there being on the east a high bank, with oak trees and pines, producing a very pretty effect. This bank, as well as the lower part of the garden, was peat or peaty sand, and as I knew this was good for rhododendrons and heaths, I was much pleased to be able to grow these plants. I did not then know, however, that this peaty soil was quite unsuited to a great many other plants, and only learnt this by the long experience which every gardener has to go through.

It was in the early part of my residence at Park-stone that I received a visit from the great French Geographer, Elisée Reclus, who had, I think, come to England to receive the gold medal of the Royal Geographical Society. He was a rather small and very delicate-looking man, highly intellectual, but very quiet in speech and manner. I really did not know that it was *he* with whose name I had been familiar for twenty years as the greatest of geographers, thinking it must have been his father or elder brother; and I was surprised when, on asking him, he said that it was himself. However, we did not talk of geography during the afternoon we spent together, but of Anarchism, of which he was one of the most convinced advocates, and I was very anxious to ascertain his exact views, which I found were really not very different from my own. We agreed that almost all social evils—all poverty, misery, and crime—were the creation of governments and of bad social systems; and that under a law of absolute justice, involving equality of opportunity and the best training for all, each local community would organize itself for mutual aid, and no great central governments would be needed, except as they grew up from the voluntary association of their parts for general and national purposes.

During the first half of my residence at Parkstone (1889–96), I did not write any new books, having, as I thought, said all that I had to say on the great subjects that chiefly interested me; but I contributed a number of articles to reviews, wrote many notices of books, with letters to *Nature* on various matters of scientific interest. A short account of the more important of these will show that I was not altogether inactive as regards literary work.

In the spring of 1890 I lectured at Sheffield and at Liverpool, and have since declined all invitations to lecture, partly from disinclination and considerations of health, but also because I believed that I could do more good with my pen than with my voice. During the year I prepared a new edition of my "Malay Archipelago," bringing the parts dealing with natural history up to date.

In the same year I contributed to the *Fortnightly Review* an article on "Human Selection," which is, I consider, though very short, the most important contribution I have made to the science of sociology and the cause of human progress. The article was written with two objects in view. The first and most important was to show that the various proposals of Grant Allen, Mr. Francis Galton, and some

American writers, to attempt the direct improvement of the human race by forms of artificial elimination and selection, are both unscientific and unnecessary; I also wished to show that the great bugbear of the opponents of social reform—too rapid increase of population—is entirely imaginary, and that the very same agencies which, under improved social conditions, will bring about a real and effective selection of the physically, mentally, and morally best, will also tend towards a diminution of the rate of increase of the population. The facts and arguments I adduce are, I believe, conclusive against the two classes of writers here referred to.

A year later I contributed a paper to the Boston *Arena*, dealing more especially with the laws of heredity and the influence of education as determining human progress, showing that such progress is at present very slow, and is due almost entirely to one mode of action of natural selection, which still eliminates *some* of the most unfit. And I pointed out that a more real and effective progress will only be made when the social environment is so greatly improved as to give to women a real choice in marriage, and thus lead both to the more rapid elimination of the lower, and more rapid increase of the higher types of humanity.

Other articles were, "A Representative House of Lords," in the *Contemporary Review* (June), and "A Suggestion to Sabbath-Keepers," in the *Nineteenth Century* (October), both which articles attracted notice in the Press. I also wrote a paper criticizing the Rev. George Henslow's view as to the origin of irregular flowers, and of spines and prickles, in *Natural Science* (September), the three articles being included in my "Studies" I also reviewed James Hutchinson Stirling's "Darwinianism" in *Nature* (February 8), and Mr. Benjamin Kidd's "Social Evolution" in the same paper (April 12), as well as an anonymous volume, entitled "Nature's Method in the Evolution of Life," by a writer who suggests vague theories, less intelligible even than those of Lucretius, as a substitute for the luminous work of Darwin.

In the next year (1895) I wrote an important article on "The Method of Organic Evolution" (*Fortnightly Review,* February-March), which was chiefly devoted to showing that the views of Mr. Francis Galton, and of Mr. Bateson in his book on "Discontinuous Variations," are erroneous; and that such variations, which are usually termed "sports," and in extreme cases "monstrosities," do *not* indicate the method of evolution.

Another article (in the October issue of the same Review) on "The Expressiveness of Speech" develops a new principle in the origin of language, and brought me a holograph (and partly unintelligible) letter from Mr. Gladstone, expressing his concurrence with it.

In July, 1895, I went with my friend and father-in-law Mr. William Mitten, for a short botanizing tour in Switzerland. We walked a good deal of the time, and I thus had a further opportunity of examining glacial phenomena. We went to Lucerne, whence we ascended the Stanzerhorn by the electric railway, and found a very interesting flora on the summit. Then to the head of the lake, and to Gœschenen, whence we walked to Andermatt; then over the Furca pass to the Rhone glacier, staying two days at the hotel; then over the Grimsel pass, where we greatly enjoyed both the flowers and the wonderful indications of glacial action, especially on the slope down to and around the Hôtel Grimsel, where we stayed the night. The valley down to Meiringen was excessively interesting, being ice-worn everywhere. We stayed an hour at the fine Handeck cascade, and then, with the help of a chaise, into which two ladies hospitably received us, got on to Meiringen. Here we stayed two days, exploring the gorge of the Aar and the wonderful rock-barrier of the Kirchet, visited the Reichenbach falls, and had an excursion to Brunig, where, in some hilly beech woods, we were greatly pleased to find the beautiful *Cephalanthere rubra* in fair numbers and in full flower. We then went on to Lauterbrunnen and the Wengern Alp, where we stayed two days, botanizing chiefly among the woods and slopes near the Trummelthel. We were, however, so dreadfully persecuted by swarms of blood-sucking flies, which filled the air and covered us in thousands, piercing through our thin clothing, that we returned home some days earlier than we had intended.

In 1896 I wrote an article on "The Gorge of the Aar and its Teachings," as serving to enforce my papers on the "Ice Age and its Work" three years before. But my most important scientific essay this year was a paper I read to the Linnæan Society on "The Problem of Utility." My purpose was to enforce the view that all specific and generic characters must be (or once have been) useful to their possessor, or, owing to the complex laws of growth, be correlated with useful characters. It was necessary to discuss this point, because Mr. Romanes had unreservedly denied it, and Professor Mivart, the Rev. Mr. Henslow, Mr. Bateson, and others, had taken the same view. I endeavoured to show that the problem is a fundamental one, that utility is the basic principle of natural selection, and that without natural selection it has not been shown how *specific* characters can arise. By *specific* is, of course, meant characters which, either separately or in combination, distinguish each species from all others, and which are found in all, or in the great bulk, of the individuals composing the species; and I have shown that it is for want of clear thinking and accurate reasoning on the entire process of species formation that the idea of useless specific characters has arisen (see "Studies," vol. i.).

During this summer I was invited by Dr. H. S. Lunn to go with him and his party to Davos for a week, early in September, and to give them a lecture on Scientific Progress in the Nineteenth Century. As I had never been in this part of Switzerland, I accepted the invitation, and had a very pleasant time. My companion on the first part of the journey was Mr. Le Gallienne, and at Basle we were joined by Dr. and Mrs. Lunn and others. At Davos we were a large party in one of the best hotels, and our special party, who sat together at meals, included the Rev. Hugh Price Hughes and the Rev. H. R. Haweis, both talented and witty men, whose presence was enough to render almost any party a brilliant success. Mr. Price Hughes, was, I think, without exception, the most witty man and one of the best companions I ever met. At breakfast and dinner he was especially amusing and brilliant, ranging from pure chaff with his old friend Dr. Lunn to genial wit and admirably narrated anecdotes. He often literally kept the table in a roar of laughter. But this was only one side of his character. He was a Christian and a humanitarian in the best sense of the words. I saw a good deal of him in private, and we often walked out together, at which times we discussed the more serious social problems of the day; and he gave me details of his rescue work in London which were in the highest degree instructive, showing that even those who are considered to be the most degraded and irreclaimable can be reached through their affections. Their degradation has usually been brought about by society, and has been intensified into hate and despair by the utterly unsympathetic and cruel treatment of our workhouses and prisons.

Dr. Lunn arranged for his party some amusement for several evenings in each week, either a concert, lecture, or conversazione. Mr. Le Gallienne gave a very interesting lecture on "English Minor Poets," reading selections from their works to illustrate their style.

My own lecture was mainly devoted to a sketch of the chief great advances of science during the century, but I added to it a kind of set-off in discoveries which had been rejected and errors which had been upheld, referring to phrenology as one of the first class, and vaccination as one of the second. There were, of course, in such a place as Davos, many doctors among the audience, and they signified their disapproval in the usual way; but I assured them that some of them would certainly live to see the time when the whole medical profession would acknowledge vaccination to be a great delusion.

Although Davos has no grand alpine scenery immediately around it, there are many delightful walks through woods full of flowers and ferns, alpine meadows with gentians and primulas, and stony passes from which the snow had just retreated. On the Strela pass, about eight thousand feet, I found some charming little alpines I had not seen before, among them the very dwarf *Viola alpina,* growing among stones, the leaves hardly visible and the comparatively large flat flowers of a very deep blue-purple, with a large orange-yellow eye.

****

On leaving Davos, I made my way across to Adelboden, where my wife and daughter, with some friends, were staying. This is surrounded with fine alpine peaks and snow-fields, and though the weather was unsettled we spent a pleasant week here— probably the last visit I shall make to ever-delightful Switzerland—the sanatorium and alpine garden of overworked Englishmen.

From this time onwards I did not write many articles or reviews, the more important being "The Problem of Instinct," in 1897, in which I gave an attempted solution of bird migration, though the article was really a review of Professor Lloyd-Morgan's "Habit and Instinct;" an article on the question whether "White Men can work in the Tropics," which most English writers declare to be impossible without thinking it necessary to adduce evidence, but which, I affirm, is proved by experience to be quite easy. Both these are reprinted in my "Studies," as is also a short essay on "The Causes of War and the Remedies," written for *L'Humanité Nouvelle.* I also wrote letters to the *Daily Chronicle* on America, Cuba, and the Philippines; and a protest against the Transvaal War in the *Manchester Guardian.*

In the year 1900 I wrote an article for the *New York Fournal* on "Social Evolution in the Twentieth Century—An Anticipation," for which I received a very complimentary letter from the editor. During the next two years I was engaged in preparing new editions of my books on "Darwinism" and "Island Life," and I also wrote several letters on political and social subjects, such as an "Appreciation of the Past Century" (in 1901, in the *Morning Leader*), and (in 1903) an article on "Anticipations and Hopes for the Immediate Future," which was written for a German paper (the *Berliner Local Anzeiger*), but which was too plain-spoken for the editor to publish, and which I accordingly sent to the *Clarion*.

In 1904 I wrote a short letter on the "Inefficiency of Strikes" for the *Labour Annual,* and a rather long one to the *Clarion,* suggesting a policy for socialists in opposition to continued military expenditure as advocated by Robert Blatchford; but this was, I fear, too much advanced even for the readers of this very advanced paper, since no one came forward in my support. I feel sure, however, that there are many who, when it is clearly put before them, will approve of the policy I have sketched out, since it is merely one of justice and consideration for nations as well as for individuals—of adopting the same rules of right and wrong in the one case as in the other.

About the year 1899 our house at Parkstone became no longer suitable owing to the fact that building had been going on all around us and what had been open country when we came there had become streets of villas, and in every direction we had to walk a mile or more to get into any open country. I therefore began to search about various parts of the southern counties for a suitable house, and after almost giving up the attempt in despair, we accidentally found a spot within four miles of our Parkstone home and about half a mile from a station, with such a charming distant view and pleasant surroundings that we determined, if we could get two or three acres at a moderate price, to build a small house upon it.

After a rather long negotiation I obtained three acres of land, partly wood, at the end of the year 1901; sold my cottage at Godalming at a fair price, began at once making a new garden and shrubbery, decided on plans, and began building early in the new year. The main charm of the site was a small neglected orchard with old much-gnarled apple, pear, and plum trees, in a little grassy hollow sloping to the south-east, with a view over moors and fields towards Poole harbour, beyond which were the Purbeck hills to the right, and a glimpse of the open sea to the left. In the foreground were clumps of gorse and broom, with some old picturesque trees, while the orchard was sheltered on both sides by patches of woodland. The house was nearly finished in about a year, and we got into it at Christmas, 1902, when we decided to call it Old Orchard.

OLD ORCHARD, BROADSTONE.
(*Built by A. R. W.,* 1902)

Being so near to our former house, I was able to bring all our choicer plants to the new ground, and there was, fortunately, a sale of the whole stock of a small nursery near Poole in the winter, at which I bought about a thousand shrubs and trees at very low prices, which enabled us at once to plant some shrubberies and flower borders, and thus to secure something like a well-stocked garden by the time we got into the house. Since that time it has been an ever-increasing pleasure, and I have been able to satisfy my craving for enjoying new forms of plant-life every year, partly by raising numbers of seeds of hardy and greenhouse plants, always trying some of the latter in sheltered places out-of-doors, and partly by exchanges or by gifts from friends, so that every year I have the great pleasure of watching the opening of some of nature's gems which were altogether new to me, or of others which increase year by year in beauty. In one end of my greenhouse I have a large warmed tank in which I grow blue, pink, and yellow water-lilies, which flower the greater part of the year, as well as a few other beautiful or curious aquatic plants, while the back wall of the house is covered with choice climbers.

In this hasty sketch of my occupations and literary work during the last nine years, I have purposely omitted the more important portion of the latter, because the circumstances that led me on to undertake three separate works, involving a considerable amount of labour, were very curious, and to me very suggestive, and I will now give a connected account of them.

When in 1896 I was invited by Dr. Lunn to give a lecture to his friends at Davos, I firmly believed that my scientific and literary work was concluded. I had been for some years in weak health, and had no expectation of living much longer. Shortly after returning from America I had a very severe attack of asthma in 1890, and a year or two after it recurred and became chronic, together with violent palpitations on the least sudden exertion, and frequent colds almost invariably followed by bronchitis. Any attempt at continuous work was therefore very far from my thoughts, though at times I was able to do a fair amount of writing.

But the very next year I obtained relief (and up to the present time an almost complete cure) in an altogether accidental way, if there are any "accidents" in our lives. Mr. A. Bruce-Joy, the well-known sculptor (a perfect stranger to me), had called on me to complete the modelling of a medallion which he had begun from photographs, and I apologized for not looking well, as I was then suffering from one of my frequent spells of asthma, which often prevented me from getting any sleep at night. He thereupon told me that if I would follow his directions I could soon cure myself. Of course, I was altogether incredulous; but when he told me that he had himself been cured of a complication of allied diseases— gout, rheumatism, and bronchitis—of many years' standing, which no English doctors were able even to alleviate, by an American physician, Dr. Salisbury; that it was effected solely by a change of diet not founded on theory or empirical treatment, but the result of thirty years' experiment on the effects of various articles of food upon men and animals, by the only scientific method of studying each food separately and exclusively, I determined to try it. The result was, that in a week I felt much better, in a month

I felt quite well, and during the ten years that have elapsed no attack of asthma or of severe palpitation has recurred, and I have been able to do my literary work as well as before I became subject to the malady.

I may say that I have long been, and am still, *in principle*, a vegetarian, and believe that, for many reasons, it will certainly be the diet of the future. But for want of adequate knowledge, and even more from the deficiencies of ordinary vegetable cookery, it often produces bad effects. Dr. Salisbury proved by experiment that it was the consumption of too much starch foods that produces the set of diseases which he especially cures; and that when these diseases have become chronic, the only cure is the almost complete abstention from starchy substances, especially potatoes, bread, and most watery vegetables, and, in place of them, to substitute the most easily digestible well-cooked meat, with fruits and nuts in moderation, and eggs, milk, etc., whenever they can be digested. Great sufferers find immediate relief from an exclusive diet of the lean of beef. I myself live upon well-cooked beef with a fair proportion of fat (which I can digest easily), a very small proportion of bread or vegetables, fruit, eggs, and light milk puddings. The curious thing is that most English doctors declare that a meat diet is to be avoided in all these diseases, and many order complete abstinence from meat, but, so far as I can learn, on no really scientific grounds, Dr. Salisbury, however, has experimentally proved that this class of ailments is due to malnutrition, and that this malnutrition is most frequently caused by the consumption of too much starch food at all meals, which overloads the stomach and prevents proper digestion and assimilation. My case and that of Mr. Bruce-Joy certainly show that Dr. Salisbury has found, for the first time in the history of medicine, a *cure*—not merely an *alleviation*—for these painful and distressing maladies. This personal detail as to my health is, I think, of general interest in view of the large number of sufferers who are pronounced incurable by English doctors, and it was here an essential preliminary to the facts I have now to relate, which would probably not have occurred as they did had my health not been so strikingly renovated.

The lecture which I gave at Davos on the science of the nineteenth century (a subject suggested by Dr. Lunn) led me to think that an instructive and popular book might be made of the subject, as I found there were so many interesting points I could not treat adequately or even refer to in a lecture. I therefore devoted most of my spare time during the next year to getting together materials and writing the volume, which I finished in the spring of 1898, and it was published in June under the title of "The Wonderful Century." At the request of my publishers I prepared from it a School Reader, with a considerable number of illustrations, which was published in 1901. This suggested the idea of a much enlarged and illustrated edition of the original work, which was, as regards many of the more important sciences and arts, a mere outline sketch. Almost all the year 1902 and part of 1903 was occupied in getting together materials for this new work, as it really was, and it was not published till the autumn of the latter year.

But while I was writing three new chapters on the wonderful astronomical progress of the latter half of the century, the startling fact was impressed upon me that we were situated very nearly at the centre of the entire stellar universe. This fact, though it had been noted by many of the greatest astronomical writers, together with many others that led to the conclusion that our universe was finite, and that we could almost, if not quite, see to its very limits, were seldom commented on as more than isolated phenomena—curiosities, as it were, of star distribution—but of no special significance. To me, however, it seemed that they probably *had* a meaning; and when I further came to examine the numerous facts which led to the conclusion that no other planet in the solar system than our earth was habitable, there flashed upon me the idea that it was only near the centre of this vast material universe that conditions prevailed rendering the development of life, culminating in man, possible. I did not, however, dwell upon this idea, but merely suggested it in a single paragraph on pp. 329–330 of my work, and I might probably never have pursued the subject further but for another circumstance which kept my attention fixed upon it.

While I was still hard at work upon this book, the London agent of the New York *Independent* wrote to ask me to write them an article on any scientific subject I chose. I at first declined, having no subject which I thought suitable, and not wishing to interrupt my work. But when he urged me again, and told me to name my own fee, the idea struck me that these astronomical facts, with the conclusion to which they seemed to me to point, might form a very interesting, and even novel and attractive article. As the subject was fresh in my mind, and I had the authorities at hand, it did not take me very long to sketch out and write a paper of the required length, which appeared simultaneously in the *Independent* and in the *Fortnightly Review*, and, to my great surprise, created quite a sensation; and, still more to my surprise, a considerable amount of antagonism and rather contemptuous criticism by astronomers and physicists, to which I replied in a subsequent article.

But as soon as my agent, Mr. Curtis Brown, read the MSS. he suggested that I should write a volume on the subject, which he was sure would be very attractive and popular, and for which he undertook to make arrangements both in England and America, and secure me liberal terms. After a little consideration I thought I could do so, and terms were arranged for the book before the article itself was published. This enabled me to get together all the necessary materials and to begin work at once, and after six months of the stiffest reading and study I ever undertook, the book, "Man's Place in the Universe," was completed in September, and published in November of the same year. In November of 1904 a cheaper edition was published, with an additional chapter in an Appendix. This chapter contained an entirely new argument, founded on the theory of organic evolution, which I had not time to introduce into the first edition. This argument is itself so powerful that, when compounded with the arguments founded on astronomical, physical, and physiological phenomena, it renders the improbability of there having been two independent developments of organic life, each culminating in man, so great as to be absolutely inconceivable. The success of this volume, and the entirely new circle of readers it brought me, caused my publishers to urge me to write my autobiography, which I should otherwise have not written at all, or only on a very much smaller scale for the information of my family as to my early life.

It seems to me a very suggestive fact that my literary work during the last ten years should have been so completely determined by two circumstances which must be considered, in the ordinary sense of the term, and in relation to my own volition, matters of chance. If Dr. Lunn had not invited me to Davos, and if he had suggested "Darwinism" or any other of my special subjects instead of the "Science of the Nineteenth Century," I should not have written my "Wonderful Century;" I should not have had my attention so specially directed to great astronomical problems; I should not, when asked for an article, have chosen the subject of our sun's central position; and I should certainly never have undertaken such a piece of work as my book on "Man's Place in the Universe," or the present autobiography. And further, without the accident of a perfect stranger calling upon me for reasons of his own, and that stranger happening to be a man who had been so marvellously cured by Dr. Salisbury as to induce me to adopt the same treatment, with similar results, I should never have had the energy required to undertake the two later and more important works. Of course, it may be that these are only examples of those "happy chances "which are not uncommon in men's lives; but, on the other hand, it may be true that, "there's a divinity that shapes our ends, rough-hew them as we will; "and those who have reason to know that spiritual beings can and do influence our thoughts and actions, will see in such directive incidents as these examples of such influence.

This concludes the narrative of "My Life" up to the date of its publication in the autumn of 1905. It will therefore be well to give here a short statement of what has occurred to me in the three succeeding years.

Ever since my friend Dr. R. Spruce died, in December, 1893, I had intended (with his executor's cordial assent) to edit so much of his journals and correspondence as related to his fourteen years of travel and residence in South America. But as time passed on each year found me so fully occupied—as here narrated—that I felt quite unable to undertake so arduous a piece of work. But when the corrected proofs of "My Life" were out of my hands, and I had no other large work in immediate prospect, I determined to begin it, especially as Spruce's old friends, Sir Clements Markham and Sir Joseph Hooker, both thought it ought to be done, and that I was the only person who could do it.

I therefore determined to begin it, and having found that Macmillan and Co. were willing to publish it, I obtained all the material—several boxes full of journals, note-books, letters, plant catalogues and descriptions, maps, and numerous partially finished sketches and drawings—from his executor, Mr. Matthew B. Slater, of Malton, Yorkshire, and also made arrangements to have all Spruce's letters to Sir William Hooker and George Bentham, Esq., which were preserved at the Kew Herbarium, carefully copied.

A large part of the journals and note-books were written in very minute script, often full of Latin names and abbreviations—a kind of hieroglyphics, as he himself termed them, which could hardly be fully utilized by any one but the writer of them. The whole would, I estimate, have given material enough to fill six or eight large volumes, and had he himself been able to devote some years of good health to the task, I have little doubt he would have produced a work which would have ranked among the classical records of travel and exploration.

All this I had to sort out, piece together, and condense into a connected narrative, occupying two volumes of moderate bulk; and to arrange for the requisite maps, and for such illustrations as could be obtained to render the work attractive. This has been, though often tedious, on the whole a labour of love for over two years. It is now going through the press, and I hope will be published very soon after the present volume.

Soon after I had begun this laborious piece of work, I was left sole executor to my dear friend (and father-in-law) Mr. William Mitten, who died in July, 1906, at the ripe age of 87. This involved me in the usual legal formalities and added to my already large correspondence, thus, to some extent, impeding my literary work. But I still continued to write articles on subjects that specially interested me. Towards the end of the same year I wrote an article on "A New House of Lords," showing how the upper chamber could become elective, and thus fulfil its true functions of a consultative and regulative body, while being generally in harmony with the best thought and opinion of the time. A second article, entitled "Personal Suffrage," laid down the principles on which a truly democratic but simple elective system might be adopted for the House of Commons. These appeared in the *Fortnightly Review* in the early months of 1907.

A little later I was asked to write one of the introductory chapters for "Harmsworth's History of the World," which I did under the title, "How Life became Possible on the Earth." In this chapter I gave a popular statement of those numerous and complex conditions and adaptations, which I had shown, in my work on "Man's Place in the Universe," to be the essential preliminary to the slow development of organic life, and which do not exist in their exact combination on any other planet.

A little earlier I had written (at request) a rather lengthy article on "Evolution and Character," intended to form part of a series of booklets on various aspects and applications of the doctrine of evolution. This series did not appear, and my article was published in the *Fortnightly Review* of January, 1908. The subject was treated in a somewhat novel way, and excited much discussion and criticism, as well as high appreciation from several unknown correspondents.

Early in 1907 I obtained a copy of Professor Percival Lowell's new and popular work, "Mars and its Canals," in which, even more confidently than before, he put forth his views as to the planet being inhabited by beings of at least the same grade of intelligence as ourselves, and that the strange "lines," forming a kind of network over its surface, really indicated irrigation canals, artificially constructed to render its deserts habitable. This whole idea seemed to me so entirely opposed to the teachings of physical and biological science, though very attractive to the public who had no means of weighing the evidence against it, and were influenced by the great reputation of the author and his extremely positive assertions, that I determined to write a reply, which occupied me for several months, and involved a large amount of labour in order to avoid errors or misconceptions. It was published by Macmillan and Co., in a small volume at the end of 1907, under the title, "Is Mars Habitable?" This question is answered by a decided negative, while much evidence is adduced showing that the strange markings, misnamed "Canals," can be sufficiently explained by an appeal to purely physical causes.

Early in the present year the discussion in Parliament of the Unemployed Workmen Bill seemed to show such a general misconception of what was required, as well as complete ignorance of experiments and methods which had been fully explained many years before, that I felt impelled to write an article pointing out the fundamental principles on which any effective remedial action must be based, and developing to some extent the detailed measures required to carry them out successfully. My proposals were founded on those of Mr. Herbert V. Mills, in his remarkable work, "Poverty and the State," published about twenty years ago, but which was overshadowed by General Booth's ineffective scheme, for which large sums of money were subscribed, while the far better plan of Mr. Mills was neglected. My article appeared in the June and July issues of the new *Socialist Review*, where I thought it might appeal most directly to those I wished especially to influence—the Labour party in the House of Commons.

This, with a few short articles and letters in English or American periodicals—discussing, among other subjects, the best mode of nationalizing railways, both in America and England—completes the record of my literary work up to the publication of the present volume.

Although I have now completed the narrative of my literary and home life, there are a number of special subjects, which, for the sake of clearness, I have either wholly omitted, or only just mentioned, but which have either formed important episodes in my life, or have brought me into communication or friendly intercourse with a number of interesting people, and which therefore require to be narrated consecutively in separate chapters. These will now follow, and will, I think, be not the least interesting or instructive portions of my work.

CHAPTER XX

LAND NATIONALIZATION TO SOCIALISM

SOON after I returned from the Amazon (about 1853), I read Herbert Spencer's "Social Statics," a work for which I had a great admiration, and which seemed to me so important in relation to political and social reform, that I thought of inviting a few friends to read and discuss it at weekly meetings. This fell through for want of support, but the whole work, and more especially the chapter on "The Right to the Use of the Earth," made a permanent impression on me, and ultimately led to my becoming, almost against my will, President of the Land Nationalization Society, which has now been over a quarter of a century in existence. In connection with this movement, I have made the acquaintance of a considerable number of persons of more or less eminence.

The publication of my "Malay Archipelago" in 1869, procured me the acquaintance of John Stuart Mill, who on reading the concluding pages, in which I condemn our "civilization" as but a form of "barbarism," and refer, among other examples, to our permitting private property in land, wrote to me from Avignon on May 19, 1870, enclosing the programme of his proposed Land Tenure Reform Association, and asking me to become a member of the General Committee. Its object was to claim the future "unearned increment" of land values for the State, to which purpose it was to be strictly limited. I accepted the offer, but proposed a new clause, giving the State power of resuming possession of any land on payment of its net value at the time, because, as I pointed out, the greatest evil was the *monopoly* of land, not the money lost by the community. This he himself supported, but suggested giving not the current value only, but something additional as compensation; and I think the clause was drawn on these lines.

The last letter I had from Mill was in April, 1871, when a great public meeting of the Association was to be held on May 3, as to which he said, "It would be very useful to the Association, and a great pleasure to myself, if you would consent to be one of the speakers at the meeting. There is the more reason why you should do so, as you are the author of one very valuable article of the programme. Were you to explain and defend that article, it would be a service which no one is so well qualified to render as yourself." I had then recently visited the stone circles and bridges of Dartmoor, and also Stonehenge, and urged the importance of preserving them. At that time there would probably have been no question of paying more than the actual selling value of the land, and we should have been spared the disgrace of having our grandest ancient monument, after centuries of neglect and deterioration, claimed to be private property, and having an exorbitant price demanded for it. But Mill's death soon afterwards put an end to the Association, and we had to wait many years for the present very imperfect legislation on the subject.

The question of land nationalization continued at intervals to occupy my mind, but having become strongly impressed by the teachings of Spencer, Mill, and other writers as to the necessity for restricting rather than extending State agency, and by their constant reference to the inevitable jobbery and favouritism that would result from placing the management of the whole land of the country in the hands of the executive, I did not attempt to write further upon the subject. But when the topic of Irish landlordism became very prominent in the year 1879–80, an idea occurred to me which seemed to entirely obviate all the practical difficulties which were constantly adduced as insuperable, and I at once took the opportunity of the controversy on the question to set forth my views in some detail. I did this especially because the Irish Land League proposed that the Government should buy out the Irish landlords, and convert their existing tenants into peasant-proprietors, who were to redeem their holdings by payments extending over thirty-five years. This seemed to me to be unsound in principle, and entirely useless except as a temporary expedient, since it would leave the whole land of Ireland in the possession of a privileged class, and would thus disinherit all the rest of the population from their native soil.

In my essay I based my whole argument upon a great principle of equity as regards the right of succession to landed property, a principle which I have since further extended to all property.[1] But the suggestion which rendered land nationalization practicable was, that while, under certain conditions stated, all land would gradually revert to the State, what is termed in Ireland the *tenant-right,* and in England the *improvements,* or increased value given to the land by the owner or his predecessors, such as buildings, drains, plantations, etc., would remain his property, and be paid for by the new state-tenants at a fair valuation. The selling value of land was thus divided into two parts: the *inherent value* or ground-rent value, which is quite independent of any expenditure by owners, but is due solely to nature and society; and the *improvements,* which are due solely to expenditure by the owners or occupiers, and which are essentially temporary in nature. My experience in surveying and land-valuation assured me that these two values can be easily separated. It follows that land as owned by the State would need no "management" whatever, the rent being merely a ground-rent, which could be collected just as the house-tax and the land-tax are collected, the state-tenant being left as completely free as is the "freeholder" now (who is in law a state-tenant), or as are the holders of perpetual feus in Scotland.

This article appeared in the *Contemporary Review* of November, 1880, and it immediately attracted the attention of Mr. A. C. Swinton, Dr. G. B. Clark, Mr. Roland Estcourt, and a few others, who had long been seeking a mode of applying Herbert Spencer's great principle of the inequity of private property in land, and who found it in the suggestions and principles I had laid down. They accordingly communicated with me; several meetings were held at the invitation of Mr. Swinton, who was the initiator of the movement, and after much discussion as to a definite programme, the "Land Nationalization Society" was formed, and, much against my wishes, I was chosen to be its president.

[1] See my "Studies, Scientific and Social," vol. ii. chap, xxviii.

Our Society being established, it seemed necessary to prepare something in the form of a handbook or introduction to the great problem of the land; and I accordingly devoted my attention to the subject, studying voluminous reports on agriculture, on Irish famines, on Highland crofters, and numbers of special treatises dealing with the various aspects of this vast and far-reaching question. My book was published in March, 1882, under the title "Land Nationalization: its

Necessity and its Aims," and gave, in a compact form, the only general account of the evils of our land system as it exists in England, Ireland, and Scotland; a comparison with other countries or places in which a better system prevails, together with a solution of the problem of how to replace it by the only just system, without any confiscation of property or injury to any living individual. The book has had a large circulation, and, in a revised edition, is still on sale; and, together with numerous tracts issued by the Society, has done much to educate public opinion on this most vital of all political or social questions.

As, however, it was quite certain that it would take a very long time before even the first steps towards land nationalization would be taken, I took every opportunity of advocating such other fundamental reforms as seemed to me demanded by equity and to be essential to social well-being. One of the earliest was on the subject of *interest*, about which there was much difference of opinion among advanced thinkers. A discussion having arisen in *The Christian Socialist*, I developed my views at some length in an article which appeared in the issue of March, 1884.

Soon after our society was started, Henry George, author of that remarkable work, "Progress and Poverty," came to England, and I had the pleasure of making his acquaintance. He spoke at several of our meetings and elsewhere in London, as well as in various parts of England and Ireland. He was a very impressive speaker, and always held his audience.

Among the most esteemed of the friends I owed to "Land Nationalization" were two eminent Scotchmen, both poets, and both ardent lovers of justice and humanity — Professor J. Stuart Blackie and Charles Mackay. The former wrote to me in July, 1882, saying that he had just finished the "careful study" of my "Land Nationalization," and that he was "happy to find it so much in accordance with my oldest and most mature speculations, and—what is of more importance—observations on the subject." He sent me a copy of his small volume, "Altavona," with a chapter on the "Sutherland Clearances," and he concluded, "As to your remedies for the gigantic evils which our present system of land laws entails, they recommend themselves strongly to every consistent thinker."

Both he and I suffered some inconvenience from having mentioned the name of the agent who carried out the terrible Sutherland evictions in the first two decades of the nineteenth century, as it is given in all the early narratives, as well as in the report of the trial of the agent for arson and murder, when, of course, he was acquitted. His sons were at that time alive, and protested against the publication. Both our publishers were frightened. Professor Blackie withdrew his book, and published a second edition much cut down. I placed mine in the hands of a new publisher, and I promised that in a new edition I would omit the name of the agent, but refused to make any alterations in the statements of facts.

Three years later (in December, 1885), when I was lecturing in Edinburgh, I had the great pleasure of meeting Professor Blackie. I was staying with the late Mr. Robert Cox, at whose house the professor was an intimate. He called soon after I arrived, and on hearing my name, he cordially embraced me (in the continental fashion) as one with whom he was in complete sympathy, and then threw himself upon the rug to talk to Mrs. Cox. Afterwards I had a long conversation with him on all the subjects that interested us most, and was delighted with his geniality no less than with his intense human sympathy, especially in the case of the cruelly disinherited Highlanders.

Although I had for many years been a great admirer of Charles Mackay's Songs and Poems, and was living quite near him while we were at Dorking, from August, 1876, to March, 1878, I did not make his acquaintance till some years afterwards, as, owing to my constitutional shyness, I do not think I ever made the first overtures to any man, or even called upon any one without some previous correspondence or introduction. But several years later I sent him a copy of my "Land Nationalization" (I think probably on the suggestion of some one who knew him), with a letter, begging his acceptance of it. This brought me three letters in rapid succession—one acknowledging it, saying he had been very ill for six months, but adding that he had been an adherent of our cause for forty years, and referring me to his poem, "Lament of Cona for the Unpeopling of the Highlands."

In the following year he removed to London for good medical attendance, and wrote me a very flattering letter after reading my "Malay Archipelago." The next year (1886) I was able to call on him, when in London for a day, at his apartments in Longridge Road, South Kensington, when we had a long talk, and he afterwards wrote to me as "My dear friend and philosopher." On the occasion of this visit he introduced me to his step-daughter, Miss Marie Corelli, a very pleasant young lady, whose future eminence as a writer I did not divine.

Notwithstanding the scanty means of the majority of the founders and members, the Land Nationalization Society has struggled on for more than a quarter of a century. Its lecturers and its yellow vans have pervaded the country, and it has effected the great work of convincing the highest and best organized among the manual workers, as represented by their Trades Unions, that the abolition of land monopoly, which is the necessary result of its private ownership, is at the very root of all social reform. Hence the future is with them and us, and though the capitalists and the official Liberals are still against us, we wait patiently, and continue to educate the masses in the certainty of a future and not distant success.

For about ten years after I first publicly advocated

land nationalization I was inclined to think that no further fundamental reforms were possible or necessary. Although I had, since my earliest youth, looked to some form of socialistic organization of society, especially in the form advocated by Robert Owen, as the ideal of the future, I was yet so much influenced by the individualistic teachings of Mill and Spencer, and the loudly proclaimed dogma, that without the constant spur of individual competition men would inevitably become idle and fall back into universal poverty, that I did not bestow much attention upon the subject, having, in fact, as much literary work on hand as I could manage. But at length, in 1889, my views were changed once for all, and I have ever since been absolutely convinced, not only that socialism is thoroughly practicable, but that it is the only form of society worthy of civilized beings, and that it alone can secure for mankind continuous mental and moral advancement, together with that true happiness which arises from the full exercise of all their faculties for the purpose of satisfying all their rational needs, desires, and aspirations.

The book that thus changed my outlook on this question was Bellamy's "Looking Backward," a work that in a few years had gone through seventeen editions in America, but had only just been republished in England. On a first reading I was captivated by the wonderfully realistic style of the work, the extreme ingenuity of the conception, the absorbing interest of the story, and the logical power with which the possibility of such a state of society as that depicted was argued and its desirability enforced. Every sneer, every objection, every argument I had ever read against socialism was here met and shown to be absolutely trivial or altogether baseless, while the inevitable results of such a social state in giving to every human being the necessaries, the comforts, the harmless luxuries, and the highest refinements and social enjoyments of life were made equally clear.

From this time I declared myself a socialist, and I made the first scientific application of my conviction in my article on "Human Selection" in the *Fortnightly Review* (September, 1890). This article called forth several expressions of approval, which I highly value. It forms the last chapter of vol. i. of my "Studies, Scientific and Social."

I now read many other books on socialism, but that which impressed me as being the most complete and thoroughly reasoned exposition, both of the philosophy and the constructive methods of socialism, was Bellamy's later work, "Equality," which comparatively few, even of English socialists, are acquainted with. The book is a sequel to "Looking Backward," and contains more than twice the matter. It shows, systematically, how our existing system of competition and individual profit — capitalism and enormous private wealth—directly lead to overwork, poverty, starvation, and crime; that it is necessarily wasteful in production and cruelly unjust in distribution; that it fosters every kind of adulteration in manufacture, and almost necessitates lying in trade; that it involves the virtual slavery of the bulk of the population, and checks or destroys any real progress of the race.

Many good people to-day who are almost horror-struck at hearing that any one they know is a socialist, would be still more amazed if they knew how many of the very salt of the earth belong (or did belong) to this despised and much-dreaded body of thinkers. Grant Allen, one of the most intellectual and many-sided men of our time, was one of us; so is Sir Oliver Lodge, one of our foremost students of physical science; and Professor Karl Pearson, a great mathematical evolutionist. Among the clergy we have the Revs. John Clifford, R. C. Fillingham, and many others among the Christian socialists, who are as much socialists as any of us. Among men of university training or of high literary ability we have H. M. Hyndman, Edward Carpenter, J. A. Hobson, Sydney Webb, Hubert Bland, H. S. Salt, J. C. Kenworthy, Morrison-Davidson, and many others. Of poets there are Gerald Massey and Sir Lewis Morris. The Labour members of Parliament are almost all socialists; while Margaret Macmillan, the Countess of Warwick, and many less-known women are earnest workers for the cause.

I may conclude this subject with the answer I recently gave to the question, "Why am I a Socialist?" I am a socialist because I believe that the highest law for mankind is justice. I therefore take for my motto, "Fiat Justitia Ruat Cœlum;" and my definition of socialism is, "The use by every one of his faculties for the common good, and the voluntary organization of labour for the equal benefit of all." That is absolute social justice; that is ideal socialism. It is, therefore, the guiding star for all true social reform.

The Vaccination Question.

I will here say a few words about another subject in which I take a great interest, and upon which I have ventured to express views contrary to those held by the orthodox authorities.

I was brought up to believe that vaccination was a scientific procedure, and that Jenner was one of the great benefactors of mankind. I was vaccinated in infancy, and before going to the Amazon I was persuaded to be vaccinated again. My children were duly vaccinated, and I never had the slightest doubt of the value of the operation—taking everything on trust without any inquiry whatever—till about 1875–80, when I first heard that there were anti-vaccinators, and read some articles on the subject. These did not much impress me, as I could not believe so many eminent men could be mistaken on such an important matter. But a little later I met Mr. William Tebb, and through him was introduced to some of the more important statistical facts bearing upon the subject.

Some of these I was able to test by reference to the original authorities, and also to the various Reports of the Registrar-General, Dr. Farr's evidence as to the diminution of small-pox *before* Jenner's time, and the extraordinary misstatements of the supporters of vaccination. Mr. Tebb supplied me with a good deal of anti-vaccination literature, especially with "Pierce's Vital Statistics," the tables in which satisfied me that the claims for vaccination were enormously exaggerated, if not altogether fallacious. I also now learnt for the first time that vaccination itself produced a disease, which was often injurious to health and sometimes fatal to life, and I also found to my astonishment that even Herbert Spencer had long ago pointed out that the first compulsory Vaccination Act had led to an increase of small-pox. I then began to study the Reports of the Registrar-General myself, and to draw out curves of small-pox mortality, and of other zymotic diseases (the only way of showing the general course of a disease as well as its annual inequalities), and then found that the course of the former disease ran so generally parallel to that of the latter as to disprove altogether any *special protective effect* of vaccination.

As I could find no short and clear statement of the main statistical facts adverse to vaccination, I wrote a short pamphlet of thirty-eight pages, entitled "Forty-five Years of Registration Statistics, proving Vaccination to be both Useless and Dangerous." This was published in 1885 at Mr.

W. Tebb's expense, and it had the effect of convincing many persons, among whom were some of my personal friends.

A few years later, when the Royal Commission on Vaccination was appointed, I was invited to become a member of it, but declined, as I could not give up the necessary time, but chiefly because I thought I could do more good as a witness. I accordingly prepared a number of large diagrams, and stated the arguments drawn from them, and in the year 1890 gave my evidence during part of three days. As about half the Commissioners were doctors, most of the others gave way to them. I told them, at the beginning of my evidence, that I knew nothing of medicine, but that, following the principle laid down by Sir John Simon and Dr. Guy, that "the evidence for the benefits of vaccination must now be statistical," I was prepared to show the bearing of the best statistics only. Yet they insisted on putting medical arguments and alleged medical facts to me, asking me how I explained this, how I accounted for that; and though I stated again and again that there were plenty of medical witnesses who would deal with those points, they continually recurred to them; and when I said I had no answer to give, not having inquired into those alleged facts, they seemed to think they had got the best of it. Yet they were so ignorant of statistics and statistical methods that one great doctor held out a diagram, showing the same facts as one of mine, and asked me almost triumphantly how it was that mine was so different. After comparing the two diagrams for a few moments I replied that they were drawn on different scales, but that with that exception I could see no substantial difference between them. The other diagram was on a greatly exaggerated vertical scale, so that the line showing each year's death-rate went up and down with tremendous peaks and chasms, while mine approximated more to a very irregular curve. But my questioner could not see this simple point; and later he recurred to it a second time, and asked me if I really meant to tell them that those two diagrams were both accurate, and when I said again that though on different scales both represented the same facts, he looked up at the ceiling with an air which plainly said, "If you will say that you will say anything!"

The Commission lingered on for six years, and did not issue its final report till 1896, while the evidence, statistics, and diagrams occupied numerous bulky blue-books. The most valuable parts of it were the appendices, containing the tables and diagrams presented by the chief witnesses, together with a large number of official tables and statistics, both of our own and foreign countries, affording a mass of material never before brought together. This enabled me to present the general statistical argument more completely and forcibly than I had done before, and I devoted several months of very hard work to doing this, and brought it out in pamphlet form in January, 1898, in order that a copy might be sent to every member of the House of Commons before the new Vaccination Act came up for discussion. This was done by the National Anti-Vaccination League, and I wrote to the half-dozen members I knew personally, begging them to give one evening to its careful perusal. But so far as any of their speeches showed, not one of the six hundred and seventy members gave even that amount of their time to obtain information on a subject involving the health, life, and personal freedom of their constituents. Yet I *know* that in no work I have written have I presented so clear and so conclusive a demonstration of the fallacy of a popular belief as is given in this work, which was entitled "Vaccination a Delusion: Its Penal Enforcement a Crime, proved by the Official Evidence in the Reports of the Royal Commission." This was included in the second part of my "Wonderful Century," published in June, 1898, and was also published separately in the pamphlet form, as it continues to be; and I feel sure that the time is not far distant when this will be held to be one of the most important and most truly scientific of my works.

CHAPTER XXI

MESMERISM TO SPIRITUALISM

I HAVE already described my first introduction to mesmerism at Leicester, how I found that I had considerable mesmeric power myself, and could produce all the chief phenomena on some of my patients; while I also satisfied myself that the almost universal opposition and misrepresentations of the medical profession were founded upon a combination of ignorance and prejudice.

During my eight years' travel in the East I heard occasionally, through the newspapers, of the strange doings of the spiritualists in America and England, some of which seemed to me too wild and *outré* to be anything but the ravings of madmen. Others, however, appeared to be so well authenticated that I could not at all understand them, but concluded, as most people do at first, that such things *must* be either imposture or delusion. How I became first acquainted with the phenomena and the effect they produced upon me are fully described in the "Notes of Personal Evidence," in my book on "Miracles and Modern Spiritualism," to which I refer my readers. I will only state here that I was so fortunate as to witness the simpler phenomena, such as rapping and tapping sounds and slight movements of a table in a friend's house, with no one present but his family and myself, and we were able to test the facts so thoroughly as to demonstrate that they were not produced by the physical action of any one of us. Afterwards, in my own house, similar phenomena were obtained scores of times, and I was able to apply tests which showed that they were not caused by any one present. A few years later I formed one of the committee of the Dialectical Society, and again witnessed, under test conditions, similar phenomena in great variety, and in these three cases, it must be remembered, no paid mediums were present, and every means that could be suggested of excluding trickery or the direct actions of any one present were resorted to.

At a later period I paid frequent visits, always with some one or more of my friends as sceptical and as earnest in search after fact as myself, to two of the best public mediums for physical phenomena I have ever met with—Mrs. Marshall and her daughter-in-law. We here made whatever investigations we pleased, and tried all kinds of tests. We always sat in full daylight in a well-lighted room, and obtained a variety of phenomena of a very startling kind, as narrated in the book referred to. During the latter part of my residence in London (1865–70) I had numerous opportunities of seeing phenomena with other mediums in various private houses in London. These were sometimes with private, sometimes with paid mediums, but always under such conditions as to render any kind of collusion or imposture altogether out of the question. During this time I was in frequent communication with Sir William Crookes, Mr. Cromwell Varley, Serjeant Cox, Mr. Hensleigh Wedgwood, and many other friends, who were either interested in or were actively investigating the subject; and through the kindness of several of them I had many opportunities of witnessing some of the more extraordinary of the phenomena under the most favourable conditions. At a much later period, when I visited America, I made the acquaintance of some of the most eminent spiritualists in Boston and Washington, and had many opportunities of seeing phenomena and obtaining tests of a different kind from any that I had seen in England; and some of these I may refer to later on.

When I had obtained in my own house the phenomena described in my "Notes of Personal Evidence," I felt sure that if any of my scientific friends could witness them they would be satisfied that they were not due to trickery, and were worthy of careful examination. I therefore endeavoured to persuade Dr. W. B. Carpenter, Professor Tyndall, and Mr. G. H. Lewes to attend *séances* and investigate the subject for themselves, but each was too incredulous to give the matter serious attention.

In 1866 I wrote a pamphlet, entitled "The Scientific Aspect of the Supernatural," which I distributed amongst my friends. After reading it, Huxley wrote that he "could not get up any interest in the subject." Tyndall read it "with deep disappointment," and he deplored my willingness to accept data unworthy of my attention.

I received many letters referring to this pamphlet, both satisfactory and otherwise, but perhaps the most interesting was that from Robert Chambers, which I here give—

"St. Andrews, February 10, 1867.
DEAR SIR,
I have received your letter of the 6th inst., and your little volume. It gratifies me much to receive a friendly communication from the Mr. Wallace of my friend Darwin's 'Origin of Species,' and my gratification is greatly heightened on finding that he is one of the few men of science who admit the verity of the phenomena of spiritualism. I have for many years *known* that these phenomena are real, as distinguished from impostures; and it is not of yesterday that I concluded they were calculated to explain much that has been doubtful in the past, and when fully accepted, revolutionize the whole frame of human opinion on many important matters.

******

How provoking it has often appeared to me that it seems so impossible, with such a man, for instance, as Huxley, to obtain a moment's patience for this subject — so infinitely transcending all those of physical science in the potential results!

431

My idea is that the term 'supernatural' is a gross mistake. We have only to enlarge our conceptions of the natural, and all will be right.

I am, dear sir,

Yours very sincerely,

ROBERT CHAMBERS."

In the latter part of the year, while attending the meeting of the British Association at Dundee, I visited St. Andrews, and after a geological excursion under the guidance of Sir A. Geikie, and a collation with the university authorities, at which Robert Chambers was present, I had the great pleasure of an hour's conversation with him in his own house.

During the years 1870–80 I had many opportunities of witnessing interesting phenomena in the houses of various friends, some of which I have not made public. Early in 1874 I was invited by John Morley, then editor of the *Fortnightly Review,* to write an article on "Spiritualism" for that periodical. Much public interest had been excited by the publication of the Report of the Committee of the Dialectical Society, and especially by Mr. Crookes's experiments with Mr. Home, and the refusal of the Royal Society to see these experiments repeated. I therefore accepted the task, and my article appeared in May and June under the title, "A Defence of Modern Spiritualism." At the end of the same year I included this article, together with my former small book,

"The Scientific Aspects of the Supernatural," and a paper I had read before the Dialectical Society in 1871, answering the arguments of Hume, Lecky, and other writers against miracles, in a volume which has had a very considerable sale, and has led many persons to investigate the subject and to become convinced of the reality of the phenomena.

The publication of my book in 1874, not only brought me an extensive correspondence on the subject, but led to my being invited to take part in many interesting *séances,* and making the acquaintance of spiritualists both at home and abroad. As what I witnessed was often very remarkable, and forms a sort of supplement to the "Notes of Personal Evidence "given in my book, and also because these phenomena have had a very important influence both on my character and my opinions, it will be necessary here to give a brief outline of them.

I attended a series of sittings with Miss Kate Cook, the sister of the Miss Florence Cook with whom Sir William Crookes obtained such very striking results. The general features of these *séances* were very similar, though there was great variety in details. They took place in the rooms of Signor Randi, a miniature-painter, living in Montague Place, W., in a large reception-room, across one corner of which a curtain was hung and a chair placed inside for the medium. There were generally six or seven persons present. Miss Cook and her mother came from North London. Miss C. was always dressed in black, with lace collar, she wore laced-up boots, and had earrings in her ears. In a few minutes after she had entered the cabinet, the curtains would be drawn apart and a white-robed female figure would appear, and sometimes come out and stand close in front of the curtain. One after another she would beckon to us to come up. We then talked together, the form in whispers; I could look closely into her face, examine the features and hair, touch her hands, and might even touch and examine her ears closely, which were *not* bored for earrings. The figure had bare feet, was somewhat taller than Miss Cook, and, though there was a general resemblance, was quite distinct in features, figure, and hair. After half an hour or more this figure would retire, close the curtains, and sometimes within a few seconds would say, "Come and look." We then opened the curtains, turned up the lamp, and Miss Cook was found in a trance in the chair, her black dress, laced-boots, etc., in the most perfect order as when she arrived, while the full-grown white-robed figure had totally disappeared.

Mr. Robert Chambers introduced me to a wealthy Scotch lady, Miss Douglas, living in South Audley Street, and at her house I attended many *séances*, and met Mr. Hensleigh Wedgwood, and several other London spiritualists. On one occasion Home was the medium and Mr. (now Sir William) Crookes was present. As I was the only one of the company who had not witnessed any of the remarkable phenomena that occurred in his presence, I was invited to go under the table while an accordion was playing, held in Home's hand, his other hand being on the table. The room was well lighted, and I distinctly saw Home's hand holding the instrument, which moved up and down and played a tune without any visible cause. On stating this, he said, "Now I will take away my hand"—which he did; but the instrument went on playing, and I saw a detached hand holding it while Home's two hands were seen above the table by all present. This was one of the ordinary phenomena, and thousands of persons have witnessed it; and when we consider that Home's *séances* almost always took place in private houses at which he was a guest, and with people absolutely above suspicion of collusion with an impostor, and also either in the daytime or in a fully illuminated room, it will be admitted that no form of legerdemain will explain what occurred. Perhaps the most interesting of these *séances* were a series with Mr. Haxby, a young man engaged in the post-office, and a remarkable medium for materializations. He was a small man, and sat in Miss Douglas's small drawing-room on the first floor separated by curtains from a larger one, where the visitors sat in a subdued light. After a few minutes, from between the curtains would appear a tall and stately East Indian figure in white robes, a rich waistband, sandals, and large turban, snowy white, and disposed with perfect elegance. Sometimes this figure would walk round the room outside the circle, would lift up a large and very heavy musical box, which he would wind up and then swing round his head with one hand. He would often come to each of us in succession, bow, and allow us to feel his hands and examine his robes. We asked him to stand against the door-post and marked his height, and on one occasion Mr. Hensleigh Wedgwood brought with him a shoe-maker's measuring-rule, and at our request, Abdullah, as he gave his name, took off a sandal, placed his foot on a chair, and allowed it to be accurately measured with the sliding-rule. After the *séance* Mr. Haxby removed his boot and had *his* foot measured by the same rule, when that of the figure was found to be full one inch and a quarter the longer, while in height it was about half a foot taller. A minute or two after Abdullah had retired into the small room, Haxby was found in a trance in his chair, while no trace of the white-robed stranger was to be seen. The door and window of the back room were securely fastened, and often secured with gummed paper, which was found intact.

On another occasion I was present in a private house when a very similar figure appeared with the medium Eglinton before a large party of spiritualists and inquirers. In this case the conditions were even more stringent and the result absolutely conclusive. A corner of the room had a curtain hung across it, enclosing a space just large enough to hold a chair for the medium. I and others examined this corner and found the walls solid and the carpet nailed down. The medium on arrival came at once into the room, and after a short period of introductions seated himself in the corner. There was a lighted gas-chandelier in the room, which was turned down so as just to permit us to see each other. The figure, beautifully robed, passed round the room, allowed himself to be touched, his robes, hands, and feet examined closely by all present—I think sixteen or eighteen persons. Every one was delighted, but to make the *séance* a test one, several of the medium's friends begged him to allow himself to be searched so that the result might be published. After some difficulty he was persuaded, and four persons were appointed to make the examination. Immediately two of these led him into a bedroom, while I and a friend who had come with me closely examined the chair, floor, and walls, and were able to declare that nothing so large as a glove had been left. We then joined the other two in the bedroom, and as Eglinton took off his clothes each article was passed through our hands, down to underclothing and socks, so that we could positively declare that not a single article besides his own clothes were found upon him. The result was published in the *Spiritualist* newspaper, certified by the names of all present.

Yet one more case of materialization may be given, because it was even more remarkable in some respects than any which have been here recorded. A Mr. Monck, a

Nonconformist clergyman, was a remarkable medium, and in order to be able to examine the phenomena carefully, and to preserve the medium from the injury often caused by repeated miscellaneous *s;eacute;ances,* four gentlemen secured his exclusive services for a year, hiring apartments for him on a first floor in Bloomsbury, and paying him a moderate salary. Mr. Hensleigh Wedgwood and Mr. Stainton Moses were two of these, and they invited me to see the phenomena that occurred. It was a bright summer afternoon, and everything happened in the full light of day. After a little conversation, Monck, who was dressed in the usual clerical black, appeared to go into a trance; then stood up a few feet in front of us, and after a little while pointed to his side, saying, "Look." We saw there a faint white patch on his coat on the left side. This grew brighter, then seemed to flicker, and extend both upwards and downwards, till very gradually it formed a cloudy pillar extending from his shoulder to his feet and close to his body. Then he shifted himself a little sideways, the cloudy figure standing still, but appearing joined to him by a cloudy band at the height at which it had first begun to form. Then, after a few minutes more, Monck again said "Look," and passed his hand through the connecting band, severing it. He and the figure then moved away from each other till they were about five or six feet apart. The figure had now assumed the appearance of a thickly draped female form, with arms and hands just visible. Monck looked towards it and again said to us "Look," and then clapped his hands. On which the figure put out her hands, clapped them as he had done, and we all distinctly heard her clap following his, but fainter. The figure then moved slowly back to him, grew fainter and shorter, and was apparently absorbed into his body as it had grown out of it.

Of course, such a narration as this, to those who know nothing of the phenomena that gradually lead up to it, seems more midsummer madness. But to those who have for years obtained positive knowledge of a great variety of facts equally strange, this is only the culminating point of a long series of phenomena, all antecedently incredible to the people who talk so confidently of the laws of nature.

Now that the whole series of similar phenomena have been co-ordinated, and to some extent rendered intelligible, by Myers's great work on "Human Personality," it is to be hoped that even students of physical science will no longer class all those who have either witnessed such phenomena or expressed their belief in them, as insane or idiotically credulous, without even attempting to show how, under similar conditions, such effects can be produced.

During my lecturing tour in the United States in 1886–87, I stayed some time in three of the centres of American spiritualism—Boston, Washington, and San Francisco, and made the acquaintance of many American spiritualists and inquirers, with whom I attended many remarkable *séances*.

At Boston I met the Rev. Minot J. Savage, whose latest work, "Can Telepathy Explain?" contains such a collection of personal experiences as have fallen to the lot of few inquirers; Mr. F. J. Garrison, a son of the great abolitionist; Mr. E. A. Brackett, a sculptor, and author of a remarkable book on "Materialized Apparitions"; Dr. Nichols, author of "Whence, Where, and Whither"; Professor James, of Harvard, and several others.

I attended several *séances* at the house of Mrs. Ross, a very good medium for materializations, in the company of one or more of ray friends. I will state what occurred on one of these occasions. The *séance* took place in a front downstairs room of a small private house, opening by sliding doors into a back room, and by an ordinary door into the passage. The cabinet was formed by cloth curtains across the corner of the room from the fireplace to the sliding door. One side of this was an outer wall, the other the wall of the back room, where there was a cupboard containing a quantity of china. I was invited to examine, and did so thoroughly—front room, floor, back room, rooms below in basement, occupied by a heating apparatus; and I am positive there were no means of communication other than the doors for even the smallest child. Then the sliding doors were closed, fastened with sticking-plaster, and privately marked with pencil. The ten visitors formed a semi-circle opposite the cabinet, and I sat with my back close to the passage door and opposite the curtain at a distance of about ten feet. A red-shaded lamp was in the furthest corner behind the visitors, which enabled me to see the time by my watch and the outlines of every one in the room; and as it was behind me the space between myself and the cabinet was very fairly lighted. Under these circumstances the appearances were as follows:—

(1) A female figure in white came out between the curtains with Mrs. Ross in black, and also a male figure, all to some distance in front of the cabinet. This was apparently to demonstrate, once for all, that, whatever they were, the figures were not Mrs. Ross in disguise.

(2) After these had retired three female figures appeared together, in white robes and of different heights. These came two or three feet in front of the curtain.

(3) A male figure came out, recognized by a gentleman present as his son.

(4) A tall Indian figure came out in white moccassins; he danced and spoke; he also shook hands with me and others, a large, strong, rough hand.

(5) A female figure with a baby stood close to the entrance of the cabinet. I went up (on invitation), felt the

baby's face, nose, and hair, and kissed it— apparently a real, soft-skinned, living baby. Other ladies and gentlemen agreed.

Directly the *séance* was over the gas was lighted, and I again examined the bare walls of the cabinet, the curtains, and the door, all being just as before, and affording no room or place for disposing of the baby alone, far less of the other figures.

At another special *séance* for friends of Dr. Nichols and Mr. Brackett, with Professor James and myself— nine in all, under the same conditions as before, eight or nine different figures came, including a tall Indian chief in war-paint and feathers, a little girl who talked and played with Miss Brackett, and a very pretty and perfectly developed girl, "Bertha," Mr. Brackett's niece, who has appeared to him with various mediums for two years, and is as well known to him as any near relative in earth-life. She speaks distinctly, which these figures rarely do, and Mr. Brackett has often seen her develop gradually from a cloudy mass, and almost instantly vanish away. But what specially interested me was, that two of the figures beckoned to me to come up to the cabinet. One was a beautifully draped female figure, who took my hand, looked at me smilingly, and on my appearing doubtful, said in a whisper that she had often met me at Miss Kate Cook's *séances* in London. She then let me feel her ears, as I had done before to prove she was not the medium. I then saw that she closely resembled the figure with whom I had often talked and joked at Signor Randi's, a fact known to no one in America.

The other figure was an old gentleman with white hair and beard, and in evening-dress. He took my hand, bowed, and looked pleased, as one meeting an old friend. Considering who was likely to come, I thought of my father and of Darwin, but there was not enough likeness to either. Then at length I recognized the likeness to a photograph I had of my cousin Algernon Wilson, whom I had not seen since we were children, but had long corresponded with, as he was an enthusiastic entomologist, living in Adelaide, where he had died not long before. Then I looked pleased and said, "Is it Algernon?" at which he nodded *earnestly*, seemed *very* much pleased, shook my hand vigorously, and patted my face and head with his other hand.

These two recognitions were to me very striking, because they were both so private and personal to myself, and could not possibly have been known to the medium or even to any of my friends present.

In Washington, where I resided several months, I made the acquaintance of Professor Elliot Coues, General Lippitt, Mr. D. Lyman, Senator and Mrs. Stanford, Mr. T. A. Bland the Indians' friend, and Mrs. Beecher Hooker, all thorough spiritualists, as well as many others unknown to fame. With the three former gentlemen I attended many *séances* of a very remarkable public medium, Mr. P. L. O. A. Keeler, and both witnessed phenomena and obtained tests of a very interesting kind. The medium was a young man of the clerk or tradesman class, with only the common school education, and with no appearance of American smartness. The arrangement of his *séances* was peculiar. The corner of a good-sized room had a black curtain across it on a stretched cord about five feet from the ground. Inside was a small table on which were a tambourine and hand-bell. Any one, before the *séances* began or afterwards, could examine this enclosed space, the curtain, the floor, and the walls. I did so myself, the room being fully lighted, and was quite satisfied that there was absolutely nothing but what appeared at first sight, and no arrangements whatever for ingress or egress but under the curtain into the room. The curtain, too, was entire from end to end, a matter of importance in regard to certain phenomena that occurred. Three chairs were placed close in front of this curtain on which sat the medium and two persons from the audience. Another black curtain was passed in front of them across their chests so as to enclose their bodies in a dark chamber, while their heads and the arms of the outer sitter were free. The medium's two hands were placed on the hands and wrist of the sitter next him.

The *séance* began with purely physical phenomena. The tambourine was rattled and played on, then a hand appeared above the curtain, and a stick was given to it which it seized. Then the tambourine was lifted high on this stick and whirled round with great rapidity, the bell being rung at the same time. All the time the medium sat quiet and impassive, and the person next him certified to his two hands being on his or hers. On one occasion a lady, a friend of Professor Elliott Coues and a woman of unusual ability and character, was the sitter, and certified at all critical times during the whole *séance* that the medium's hands were felt by her. After these and many other things were performed, the hand would appear above the curtain, the fingers moving excitedly. This was the signal for a pencil and a pad of note-paper (as commonly used in America); then rapid writing was heard, a slip of paper torn off and thrown over the curtain, sometimes two or three in rapid succession, and in the direction of certain sitters. The director of the *séance* picked them up, read the name signed, and asked if any one knew it, and when claimed it was handed to him. In this way a dozen or more of the chance visitors received messages which were always intelligible to them and often strikingly appropriate. I will give some of the messages I thus received myself.

On my second visit a very sceptical friend went with us, and seeing the writing-pad on the piano marked several of the sheets with his initials. The medium was very angry and said it would spoil the *séance*. However, he was calmed by his friends. When it came to the writing the pad was given to me over the top of the curtain to hold. I held it just above the medium's shoulder, when a hand and pencil came *through the curtain*, and wrote on the pad as I held it. It is a bold scrawl and hard to read, but the first words seem to be, "Friends were here to write, but only this one could. ... A. W." Another evening, with the same medium, I received a paper with this message, "I am William Martin, and I come for Mr. William Wallace, who could not write this time after all. He wishes to say to you that you shall be sustained by coming results in the position you have taken in the Ross case. It was a most foul misrepresentation."

This, and other writing I had afterwards, are to me striking tests in the name William Martin. I never knew him, but he was an early friend of my brother who was for some time with Martin's father to learn practical building, the latter being then engaged in erecting King's College. When I was with my brother learning surveying, etc., he used often to speak of his friend Martin, but for the last forty-five years I had never thought of the name and was greatly surprised when it appeared. About a month later I had the following message from the elder Martin, written in a different hand:—

"MR WALLACE,
Your father was an esteemed friend, and I like to come to

you for his sake. We are often together. How strange it seems to us here that the masses can so long exist in ignorance. Console yourself with the thought that though ignorance, superstition and bigotry have withheld from you the just rewards to which your keen enlightenment and noble sacrifices so fully entitle you, the end is not yet, and a mighty change is about to take place to put you where you belong.

WILLIAM MARTIN."

I have no evidence that this Mr. Martin was a friend of my father, but the fact that my brother William was with him as stated (which must have been a favour), renders it probable. On the same evening there was a number of messages to about a dozen people all in different handwritings, several of which were recognized. My friend General Lippitt had a most beautiful message which he allowed me to copy, as it was a wonderful test and greatly surprised and delighted him. His first wife had died twenty-seven years before in California. She was an English lady and he was greatly attached to her. This is the message:—

"DARLING FRANCIS,

I come now to greet you from the high spheres to which I have ascended. Do you recall the past? Do you remember this day? This day I used to look forward to and mention with such pride? This, my darling, is my birthday anniversary.

447

Do you not remember? Oh how happy shall we be when reunited in a world where we shall see as we are seen and know as we are known.

ELIZABETH LIPPITT."

General Lippitt told me it *was* his first wife's birthday, that he had not recollected it that day, and that no one in Washington knew the fact but himself.

A German gentleman who was present had a message given him, which was not only written, as he declared, in excellent German, but was very characteristic of the friend from whom it purported to come.

On this evening most wonderful physical manifestations occurred. A stick was pushed out *through* the curtain. Two watches were handed to me *through* the curtain, and were claimed by the two persons who sat by the medium. The small tambourine, about ten inches diameter, was pushed *through* the curtain and fell on the floor. These objects came through different parts of the curtain, but left no holes as could be seen at the time, and was proved by a close examination afterwards. More marvellous still (if that be possible), a waistcoat was handed to me over the curtain, which proved to be the medium's, though his coat was left on and his hands had been held by his companion all the time; also about a score of people were looking on all the time in a well-lighted room. These things *seem* impossible, but they are, nevertheless, facts.

At San Francisco my time was short, and my experiences were limited to a slate-writing *séance* of a striking and very satisfactory nature. I went with my brother John who had lived in California nearly forty years, and who, the day before, had bought a folding-slate bound with list to shut noiselessly. The *séance* was in the morning of a bright sunny day, and we sat at a small table close to a window. Mr. Owen, the editor of the *Golden Gate,* with a friend (a physician), accompanied us; but they sat a little way from the table, looking on. The medium, Mr. Fred Evans, was quite a young man, whose remarkable gift had been developed under Mr. Owen's guidance.

From a pile of small slates on a side-table four were taken at a time, cleaned with a damp sponge, and handed to us to examine, then laid in pairs on the table. All our hands were then placed over them till the signal was given, and on ourselves opening them writing was found on both slates. Two other pairs were then similarly placed on the table, on one of which the medium drew two diagonal pencil lines, and on that slate writing was produced in five different colours—deep blue, red, light green-blue, pale red-lilac, deep lilac, and these could be seen all superposed upon the pencil cross-lines. My brother's folding-slate was then placed upon the floor a foot or two away from the table, and after we had conversed for a few minutes, keeping it in sight, it was found to be written on both the inner sides. It then occurred to me to ask the medium whether writing could be produced on paper placed between slates. After a moment's pause, as if asking the question of his guides, he told me to take a paper pad, tear off six pieces, and place them all between a pair of slates. This I did, and we placed our hands over them as before, and in a few minutes, on opening them, we found six portraits in a peculiar kind of crayon drawing.

I will now describe the writings and drawings we obtained, which are now before me. The first was a letter filling the slate in small, clear, and delicate writing, of which I will quote the concluding portion: "I wish I could describe to you my spirit home. But I cannot find words suitable in your earthly language to give it the expression it deserves. But you will know all when you join me in the spirit world.... Your loving sister, Elizabeth Wallace. Herbert is here."

Here are two family names given, the first being one which no one else present could have known, as she died when we were both schoolboys. The opening and concluding parts of the letter show that it was addressed specially to myself. The next was addressed to my brother, referring to me as "brother Alf," and is signed "P. Wallace." This we cannot understand, as we have no relative with that initial, except a cousin, Percy Wilson. It is, I think, not improbable that in transferring the message through the medium, and perhaps through a spirit-scribe (as is often said to be the case), the surname was misunderstood owing to the latter supposing that the communicant was a brother.

The next slate contains a message signed "Judge Edmonds," addressed to myself and Mr. Owen, on the general subject of spirit manifestations. It is written very distinctly in a flowing hand.

The next is the slate written in five colours, and signed "John Gray," one of the well-known early advocates of spiritualism in America. It is also on the general subject of spirit-return. Then comes a slate containing a portrait and signature of "Jno. Pierpont," one of the pioneers of spiritualism, and around the margin three messages in different handwritings. One is from Stanley St. Clair, the spirit-artist, who says he has produced the portrait for me, at the request of the medium. The others are short messages from Elizabeth Wallace and R. Wallace, the latter perhaps one of the unknown Scotch uncles of my father, the other beginning, "God bless you, my boys," is probably from our paternal grandmother, who is buried at Laleham. The last is my brother's folding-slate, containing on one side a short farewell from "John Gray," the signature being written three times in different styles and tints; the other side is a message signed, "Your father, T. V. Wallace." This, again, was a test, as no one present would have been able to give my father's unusual initials correctly, and as he was accustomed to sign his name.

The six portraits on paper with the lips tinted are those of Jno. Pierpont (signed); Benjamin Rush (an early spiritualist, signed); Robt. Hare, M.D., whose works I had quoted (signed); D. D. Home, the celebrated medium who had died the year before—a likeness easily recognized; a girl (signed "The Spirit of Mary Wallace "), probably my sister who had died the year before I was born, when eight years old; and a lady, who was recognized as Mrs. Breed, a medium of San Francisco. These are all rather rude outlines, in somewhat irregular and interrupted dashes, but they are all lifelike, and considering that they must have been precipitated on the six surfaces while in contact with each other between the slates, as placed by myself, are exceedingly curious. The whole of these seven plates and six papers were produced so rapidly that the *séance* occupied less than an hour, and with such simple and complete openness, under the eyes of four observers, as to constitute absolutely test conditions, although without any of the usual paraphernalia of tests which were here quite unnecessary. A statement to this effect was published, with an account of the *séance,* signed by all present.

During the last fifteen years I have not seen much of spiritualistic phenomena; but those who have read the account of my early investigations in my book on the subject, and add to them all that I have indicated here, will see that I have reached my present stand-point by a long series of experiences under such varied and peculiar conditions as to render unbelief impossible. As Dr. W. B. Carpenter well remarked many years ago, people can only believe new and extraordinary facts if there is a place for them in their existing "fabric of thought." The majority of people to-day have been brought up in the belief that miracles, ghosts, and the whole series of strange phenomena here described cannot exist; that they are contrary to the laws of nature; that they are the superstitions of a bygone age; and that therefore they are necessarily either impostures or delusions. There is no place in the fabric of their thought into which such facts can be fitted. When I first began this inquiry it was the same with myself. The facts did not fit into my then existing fabric of thought. All my preconceptions, all my knowledge, all my belief in the supremacy of science and of natural law were against the possibility of such phenomena. And even when, one by one, the facts were forced upon me without possibility of escape from them, still, as Sir David Brewster declared after being at first astounded by the phenomena he saw with Mr. Home, "spirit was the last thing I could give in to." Every other possible solution was tried and rejected. Unknown laws of nature were found to be of no avail when there was always an unknown intelligence behind the phenomena—an intelligence that showed a human character and individuality, and an individuality which almost invariably *claimed* to be that of some person who had lived on earth, and who, in many cases, was able to prove his or her identity. Thus, little by little, a place was made in my fabric of thought, first for all such well-attested facts, and then, but more slowly, for the spiritualistic interpretation of them.

Unfortunately, at the present day most inquirers begin at the wrong end. They want to see, and sometimes do see the most wonderful phenomena first, and being utterly unable to accept them as facts denounce them as impostures, as did Tyndall and G. H. Lewes, or declare, as did Huxley, that such phenomena do not interest them. Many people think that when I and others publish accounts of such phenomena, we wish or require our readers to believe them on *Our* testimony. But that is not the case. Neither I nor any other well-instructed spiritualist expects anything of the kind. We write not to convince, but to excite to inquiry. We ask our readers not for *belief,* but for doubt of their own infallibility on this question; we ask for inquiry and patient experiment before hastily concluding that we are, all of us, mere dupes and idiots as regards a subject to which we have devoted our best mental faculties and powers of observation for many years.

CHAPTER XXII

A CHAPTER ON MONEY MATTERS— EARNINGS AND LOSSES—SPECULATIONS AND LAW-SUITS

UP to the age of twenty-one I do not think I ever had a sovereign of my own. I then received a small sum, perhaps about £50, the remnant of a legacy from my grandfather, John Greenell. This enabled me to get a fair outfit of clothes, and to keep myself till I got the appointment at the Leicester school. While living at Neath as a surveyor I did little more than earn my living, except during the six months of the railway mania, when I was able to save about £100. This enabled me to go to Para with Bates, and during the four years on the Amazon my collections just paid all expenses, but those I was bringing home with me would probably have sold for £200. My agent, Mr. Stevens, had fortunately insured them for £150, which enabled me to live a year in London, and get a good outfit and a sufficient cash balance for my Malayan journey.

My eight years in the Malay Archipelago were successful, financially, beyond my expectations. Celebes, the Moluccas, the Aru Islands, and New Guinea were, for English museums and private collections, an almost unknown territory. A large proportion of my insects and birds were either wholly new or of extreme rarity in England; and as many of them were of large size and of great beauty, they brought very high prices. My agent had invested the proceeds from time to time in Indian guaranteed railway stock, and a year after my return I found myself in possession of about £300 a year. Besides this, I still possessed the whole series of private collections, including large numbers of new or very rare species, which, after I had made what use of them was needed for my work, produced an amount which in the same securities would have produced about £200 a year more.

But I never reached that comfortable position. Owing to my never before having had more than enough to supply my immediate wants, I was wholly ignorant of the numerous snares and pitfalls that beset the ignorant investor, and I unfortunately came under the influence of two or three men who, quite unintentionally, led me into trouble. Soon after I came home I made the acquaintance of Mr. R., who held a good appointment under Government, and had, besides, the expectation of a moderate fortune on the death of an uncle. I soon became intimate with him, and we were for some years joint investigators of spiritualistic phenomena. He was, like myself at that time, an agnostic, well educated, and of a more positive character than myself. He had for some years saved part of his income, and invested it in various foreign securities at low prices, selling out when they rose in value, and in this way he assured me he had in a few years doubled the amount he had saved. He studied price-lists and foreign news, and assured me that it was quite easy, with a little care and judgment, to increase your capital in this way. He quite laughed at the idea of allowing several thousand pounds to lay idle, as he termed it, in Indian securities, and so imbued me with an idea of his great knowledge of the money market, that I was persuaded to sell out some of my bonds and debentures and buy others that he recommended, which brought in a higher interest, and which he believed would soon rise considerably in value. This change went on slowly with various success for several years, till at last I had investments in various English, American, and foreign railways, whose fluctuations in value I was quite unable to comprehend, and I began to find, when too late, that almost all my changes of investment brought me loss instead of profit; and later on, when the great depression of trade of 1875–85 occurred, the loss was so great as to be almost ruin.

In 1866 one of my oldest friends became secretary to a small body of speculators, who had offices in Pall Mall, and who, among other things, were buying slate quarry properties, and forming companies to work them.

I was persuaded to take shares, and to be a director of these companies, without any knowledge of the business, or any idea how much capital would be required. The quarries were started, machinery purchased, call after call made, with the result in both cases that, after four or five years of struggle, the capital required and the working expenses were so great that the companies had to be wound up, and I was the loser of about a thousand pounds.

While this was going on a still more unfortunate influence became active. My old friend in Timor and Singapore, Mr. Frederick Geach, the mining engineer, came home from the East, and we became very intimate, and saw a good deal of each other. He was a Cornishman, and familiar with tin, lead, and copper mining all his life, and he had the most unbounded confidence in good English mines as an investment. He had shares in some of the lead-mines of Shropshire and Montgomeryshire, and we went for a walking tour in that beautiful country, visited the mines, went down the shafts by endless perpendicular ladders, and examined the veins and workings with the manager, who had great confidence in its value, and was a large shareholder. "Here," said Geach, "you can see the vein of lead ore. It is very valuable, and extends to an unknown depth. This is not a probability, it is a certainty." And so I was persuaded to buy shares in lead-mines, and gradually had a large portion of my capital invested in them. About 1870 the price of lead began to fall, and has continued to fall ever since. The result of all this was that by 1880 a large part of the money I had earned at the risk of health and life was irrecoverably lost.

While these continued misfortunes were in progress I was involved in two other annoyances, causing anxiety and worry for years, as well as a very large money loss. The first was with a dishonest builder, who contracted to build my house at Grays, and who was paid every month according to the proportion of the work done. One day, when the house was little more than half finished, he did not appear to pay his men, and as they would not continue to work without their money I paid them. He did not appear the next week, and sent no excuse, so the architect gave him notice that I should complete the building myself, and that, according to the agreement, he would be responsible for any cost beyond the contract price. After a few weeks he appeared, and wanted to go on, but that we declined. The house cost me somewhat more than the contract price, and when it was finished I sent him word he could have his ladders, scaffold-poles, boards, etc., though, according to the agreement, they were to be my property on his failure to finish the building.

I soon found, however, that he had not paid for a large portion of the materials, and bills kept coming in for months afterwards for bricks, timber, stone, iron-work, etc., etc. The merchants who had trusted him found that he had no effects whatever, as he lived as a lodger with his father; and from all I heard, was accustomed to take contracts in different places round London, and by not paying for any materials that he could get on credit, made a handsome profit. But the height of his impudence was to come. About five years after the house was finished, I received a demand through a lawyer for (I think) between £800 and £900 damages for not allowing this man to finish the house! I wrote, refusing to pay a penny. Then came a notice of an action at law; and I was obliged to put it in a lawyer's hands. All the usual preliminaries of interrogatories, affidavits, statements of claim, replies, objections, etc., etc., were gone through, and on every point argued we were successful, with costs, which we never got. The case was lengthened out for two or three years, and then ceased, the result being that I had to pay about £100 law costs for what was merely an attempt to extort money. That was my experience of English *law,* which leaves the honest man in the power of the dishonest one, mulcts the former in heavy expenses, and is thus the very antithesis of *justice.*

The next matter was a much more serious one, and cost me fifteen years of continued worry, litigation, and persecution, with the final loss of several hundred pounds. And it was all brought upon me by my ignorance and my own fault—ignorance of the fact so well shown by the late Professor de Morgan —that "paradoxers," as he termed them, can never be convinced, and my fault in consenting to get money by any kind of wager. It constitutes, therefore, the most regrettable incident in my life. As many inaccurate accounts have been published, I will now state the facts, as briefly as possible, from documents still in my possession.

In *Scientific Opinion* of January 12, 1870, Mr. John Hampden (a relative of Bishop Hampden) challenged scientific men to prove the convexity of the surface of any inland water, offering to stake £500 on the result. It contained the following words: "He will acknowledge that he has forfeited his deposit if his opponent can exhibit, to the satisfaction of any intelligent referee, a convex railway, river, canal, or lake." Before accepting this challenge I showed it to Sir Charles Lyell, and asked him whether he thought I might accept it. He replied, "Certainly. It may stop these foolish people to have it plainly shown them." I therefore wrote accepting the offer, proposing Bala lake, in North Wales, for the experiment, and Mr. J. H. Walsh, editor of the *Field,* or any other suitable person, as referee. Mr. Hampden proposed the Old Bedford canal in Norfolk, which, near Downham Market, has a stretch of six miles quite straight between two bridges. He also proposed a Mr. William Carpenter (a journeyman printer, who had written a book upholding the "flat earth" theory) as his referee; and as Mr. Walsh could not stay away from London more that one day, which was foggy, I chose Mr. Coulcher, a surgeon and amateur astronomer, of Downham Market, to act on my behalf, Mr. Walsh being the umpire and referee.

The experiment finally agreed upon was as follows: The

iron parapet of Welney bridge was thirteen feet three inches above the water of the canal. The Old Bedford bridge, about six miles off, was of brick and somewhat higher. On this bridge I fixed a large sheet of white calico, six feet long and three feet deep, with a thick black band along the centre, the lower edge of which was the same height from the water as the parapet of Welney bridge; so that the centre of it would be as high as the line of sight of the large six-inch telescope I had brought with me. At the central point, about three miles from each bridge, I fixed up a long pole with two red discs on it, the upper one having its centre the same height above the water as the centre of the black band and of the telescope, while the second disc was four feet lower down. It is evident that if the surface of the water is a perfectly straight line for the six miles, then the three objects—the telescope, the top disc, and the black band—being all exactly the same height above the water, the disc would be seen in the telescope projected upon the black band; whereas, if the six-mile surface of the water is convexly curved, then the top disc would appear to be decidedly higher than the black band, the amount due to the known size of the earth being five feet eight inches, which amount will be reduced a little by refraction to perhaps about five feet.

The following diagrams illustrate the experiment made. The curved line in Fig. 1, and the straight line in Fig. 2, show the surface of the canal on the two theories of a round or a flat earth. A and C are the two bridges six miles apart, while B is the pole midway with two discs on it, the upper disc, the telescope at A, and the black line on the bridge at C, being all exactly the same height above the water. If the surface of the water is truly flat, then on looking at the mark C with the telescope A, the top disc B will cover that mark. But if the surface of the water is curved, then the upper disc will appear above the black mark, and if the disc is more than four feet above the line joining the telescope and the black mark, then the lower disc will also appear above the black mark. Before the experiment was made a diagram similar to this was submitted to Mr. Hampden, his referee Mr. Carpenter, and Mr. Walsh, and all three agreed that it showed clearly what should be seen in the two cases, while the former declared their firm belief that Fig. 2 showed what *would* be seen.

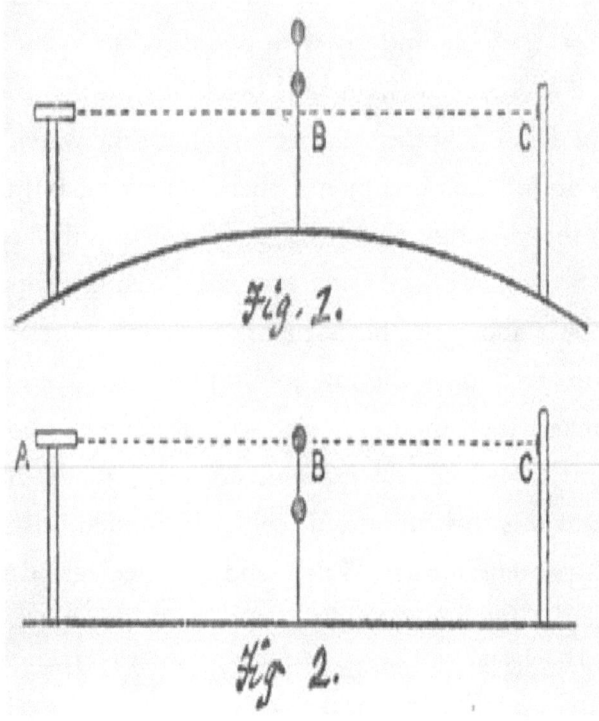

When the pole was set up and the mark put upon the bridge, Mr. Carpenter accompanied me, and saw that their heights above the water were the same as that of the telescope resting on the parapet of the bridge. What was seen in the large telescope was sketched by Mr. Coulcher and signed by Mr. Carpenter as correct, and is shown in the following diagram which was reproduced in the *Field* newspaper (March 26, 1870), and also in a pamphlet by Carpenter himself. But he declared that this proved nothing, because the telescope was not levelled, and because it had no cross-hair!

At his request to have a spirit-level in order to show if there was any "fall" of the surface of water,

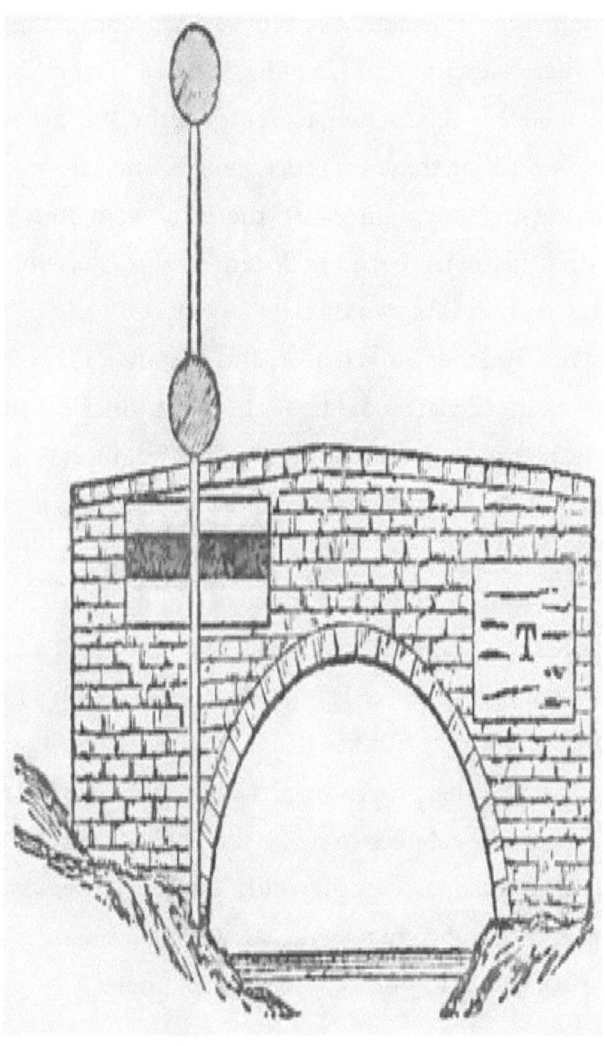

"Signed by Mr. Carpenter."—*Dr. Coulcher's Report.* *"Signed!"*

I had been to King's Lynn and borrowed a good Troughton's level from a surveyor there. This I now set up on the bridge at exactly the same height above the water as the other telescope, and having levelled it very accurately and called Mr. Carpenter to see that the bubble was truly central and that the least movement of the screws elevating or depressing it would cause the bubble to move away, I adjusted the focus on to the distant bridge, showing also the central staff and its two discs.

Mr. Coulcher looked at it, and then Mr. Carpenter, and the moment the latter did he said, "Beautiful! Beautiful!" and on Mr. Hampden asking him if it was all right, he replied that it was perfect, and that it showed the three points in "a perfect straight line;" "as level as possible!" And he actually jumped for joy. Then I asked Mr. Coulcher and Mr. Carpenter both to make sketches, which they did. We then fixed a calico flag on the parapet to make it more visible, and drove back with the instruments to Old Bedford bridge, where I set up the level again at the proper height above the water, and again asked both the referees to make sketches of what was seen in the level-telescope. This they did. Mr. Carpenter's was rather more accurately drawn, and Mr. Coulcher signed them as being correct, and both are reproduced here.

THE "BEDFORD LEVEL" SURVEY.—SKETCHES BY THE TWO REFEREES.

Copied from the *Field* for March 26, 1870.

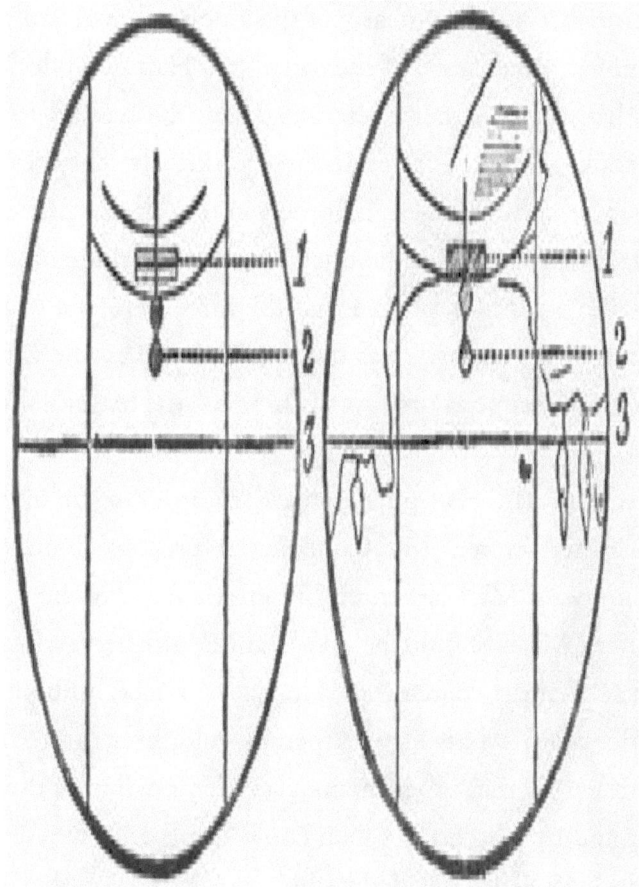

These two views, as seen by means of the *inverting* telescope, are exact representations of the sketches taken by Mr. Hampden's Referee, and attested by Dr. Coulcher as being correct in both cases: first, from Welney Bridge; and secondly, from the Old Bedford Bridge.

The view in the large telescope and in the level-telescope both told exactly the same thing, and, moreover, proved that the curvature was very nearly of the amount calculated from the known dimensions of the earth. Mr. Hampden declined to look through either telescope, saying he trusted to Mr. Carpenter; while the latter declared positively that they had won, and that we knew it; that the fact that the distant signal *appeared* below the middle one as far as the middle one did below the cross-hair, proved that the three were in a straight line, and that the earth was flat, and he rejected the view in the large telescope as proving nothing for the reasons already stated.

At first Mr. Hampden refused to appoint an umpire, because my referee, Mr. Coulcher, refused to discuss the question with Mr. Carpenter; but after a few days he agreed that Mr. Walsh should be the umpire, after receiving the reports of the two referees. He had, in fact, unbounded confidence in what Mr. Carpenter told him, and firmly believed that the experiments had demonstrated the flat earth, and that no honest man could think otherwise.

But Mr. Walsh decided without any hesitation that I had proved what I undertook to prove. He published the whole of the particulars with the reports of the referees and their sketches in the *Field* of March 18 and 26, while a considerable correspondence and discussion went on for some weeks later.

At Mr. Hampden's request he allowed Mr. Carpenter to send in a long argument to show that the experiments were all in Mr. Hampden's favour, and having considered them, he wrote to Mr. Hampden that he should hand me the stakes on a certain day if he had no other reason to adduce why he should not do so. Thereupon Mr. Hampden wrote to him *demanding his money back* on the ground that the decision was unjust, and ought to have been given in his favour.

In thus writing to Hampden and receiving his demand for his deposit to be returned, Mr. Walsh made a great mistake, which had serious consequences for me. The law declares that all wagers are null and void, and that money lost by betting is not recoverable at law. But the judges have decided that when a wager is given against him by the umpire, the loser can claim his money back from the stakeholder if the latter has not already paid it away to the winner. Hence, if a loser immediately claims his money from the stake-holder, the law will enforce the former's claim on the ground that it is *his* money, and the fact that he has lost it in a quite fair wager is beyond the cognizance of the law. Neither I nor Mr. Walsh knew of this, although he had decided and paid many wagers; but this resulted in my having to pay the money back five years later, as will be presently described.

I will now briefly state what were Hampden's proceedings for the next fifteen or sixteen years. He first began abusing Mr. Walsh in letters, postcards, leaflets, and pamphlets, as a liar, thief, and swindler. Then he began upon me with even more virulence, writing to the presidents and secretaries of all the societies to which I belonged, and to any of my friends whose addresses he could obtain. One of his favourite statements in these letters was, "Do you know that Mr. A. R. Wallace is allowing himself to be posted all over England as a cheat and a swindler?" But he soon took more violent measures, and sent the following letter to my wife:—

"MRS. WALLACE,

Madam—If your infernal thief of a husband is brought home some day on a hurdle, with every bone in his head smashed to pulp, you will know the reason. Do you tell him from me he is a lying infernal thief, and as sure as his name is Wallace he never dies in his bed.

You must be a miserable wretch to be obliged to live with a convicted felon. Do not think or let him think I have done with him.

JOHN HAMPDEN."

For this I brought him up before a police magistrate, and he was bound over to keep the peace for three months, suffering a week's imprisonment before he could find the necessary sureties. But as soon as the three months were up, he began again with more abuse than ever, distributing tracts and writing to small local papers all over England.

I now began to receive letters from friends, and also from perfect strangers, asking me if I knew what was said about me everywhere. I will give a summary of the steps I was obliged to take with the results, or rather absence of results, that followed.

In 1871, Mr. Walsh prosecuted Hampden for libel. He was convicted at the Old Bailey, and bound over to keep the peace for one year.

In January, 1871, I brought an action for libel in order to give Hampden the opportunity of justifying, if he could, his language towards me. He did not defend the action, but suffered judgment to go by default, and the jury gave me a verdict with £600 damages. But whatever property he had had been transferred to his son-in-law (a solicitor), so I could not get a penny, and had to pay the costs of the suit which, though undefended, were heavy.

In October, 1872, I prosecuted him at the Old Bailey for further libels. He was respited on publicly apologizing in several newspapers.

Some months afterwards, however, he began again with equally foul libels, and I had him brought up under his recognizances, when he was sentenced to two months' imprisonment in Newgate.

But within a year he began again as violently as ever, and on March 6, 1875, he was indicted at Chelmsford Assizes for fresh libels, and on proof of his previous convictions and apologies, he was sentenced to one year's imprisonment and to keep the peace, under heavy recognizances and sureties, for two years more. (A full report is given in the *Chelmsford Chronicle,* March 12, 1875.)

Through the interest of his friends, however, he was liberated in about six months; and thereupon, in January, 1876, he brought an action against Mr. Walsh to recover his deposit of £500, and this action he won, on the grounds already stated; and as I had signed an indemnity to Mr. Walsh, I had to pay back the money, and also pay all the costs of the action, about £200 more. But as I had a judgment for £687 damages and costs in my libel suit against Hampden, I transferred this claim to Mr. Walsh as a set-off against the amount due by him. Hampden, however, had already made himself a bankrupt to prevent this claim being enforced, and had assigned all his actual or future assets to his son-in-law.

There were now legal difficulties on both sides. I was advised that the bankruptcy was *fraudulent,* and could be annulled; but to attempt this would be costly, and the result uncertain. On the other hand, it was doubtful whether my claim against Hampden would not be treated as an ordinary creditor's claim in the bankruptcy. There was, therefore, a consultation of the solicitors, and a voluntary arrangement was arrived at. I was to pay all the costs of the suit and £120, amounting to £277; while £410 still remained nominally due to me from Hampden.

These terms were formally agreed to by Hampden and his son-in-law, and were duly carried out. Of course I had also to pay Mr. Walsh's costs in the action and my own lawyer's bill for the settlement, as well as those of the action for libel, and the various criminal prosecutions of Hampden I had been compelled to undertake.

About this time he printed one thousand copies of a two-page leaflet, and sent them to almost every one in my neighbourhood whose address he could obtain, including most of the masters of Charterhouse School, and the residents as well as the tradesmen of Godalming. It was full of—"scientific villainy and roguery,"—"cheat, swindler, and impostor."—"My specific charge against Mr. A. R. Wallace is that he obtained possession of a cheque for £1000 by fraud and falsehood of a party who had no authority to dispose of it." To save trouble, I drew up a short circular stating the main facts already given here for the information of those who had received Hampden's absurdly false libels, and thereafter took no further notice of him.

He continued to circulate his postcards and tracts, and to write to all manner of people, challenging them to prove that the earth was not flat, for several years after. The last of his efforts which I have preserved is an eight-page tract, which he distributed at the Royal Geographical Society's Exhibition of Geographical Appliances, in December, 1885, in which he attacks all geographical teaching in his usual style, and declares that "at the present moment they are cowering beneath the inquiring gaze of one single truth-seeker, JOHN HAMPDEN, the well-known champion of the Mosaic cosmogony, as against the infidel theories and superstitions of the pagan mystics, who is, at the end of fifteen years' conflict, still holding his ground against all the professional authorities of England and America; and the single fact that during the whole of that time, no one but a degraded swindler has dared to make a fraudulent attempt to support the globular theory, is ample and overwhelming proofs of the worthless character of modern elementary geography." And this man was educated at Oxford University! Seldom has so much boldness of assertion and force of invective been combined with such gross ignorance. And to this day a society exists to uphold the views of Hampden, Carpenter, and their teacher, "Parallax!"

The two law suits, the four prosecutions for libel, the payments and costs of the settlement, amounted to considerably more than the £500 I received from Hampden, besides which I bore all the costs of the week's experiments, and between fifteen and twenty years of continued persecution—a tolerably severe punishment for what I did not at the time recognize as an ethical lapse.

There is one other small money matter which I wish to put on record here, because, though it involves only the small sum of sixpence, it affords an example of official meanness, and what really amounts to petty larceny, which can hardly be surpassed. In 1865 the British Museum purchased from me some specimen (I think a skeleton) for which they agreed to pay £5. Two years later I received the following printed form:—

"Principal Librarian and Secretary's Office.
British Museum, W.C., June 24, 1867.

SIR,
If you will send your own stamped receipt to this Office, you will be paid the amount due to you by the Trustees of the British Museum, £5 0s. 0d.
I am, sir,
Your very obedient Servant,
THOMAS BUTLER,
Assist. Secretary.
Mr. A. R. Wallace."

I, of course, complied with the request and sent the stamped receipt, and by return of post had the following written communication:—

"Mr. Butler begs to transmit the enclosed P.O. order for £4 19s. 6d. to Mr. Wallace, and the amount of it, with the cost of the order (6d.), makes up the sum due by the Trustees to Mr. Wallace.

    British Museum, June 25, 1867."

This amazing little dodge (for I can call it nothing else) completely staggered me. I was at first inclined to return the P.O. order, or to write asking for the 6*d.*, and if necessary summon Mr. Butler (or the Trustees) to a County Court for the 6*d.* due. But I was busy, and did not want to enter upon what I felt sure would be a long correspondence and endless trouble and expense. I therefore determined to keep the two incriminating documents, and some day print them. The day has now come; and it may be interesting to learn whether this preposterous and utterly dishonest method of paying part of an admitted debt, after obtaining a receipt for the whole, continues to be practised in this or any other public institution.

    It was while these troubles in the Hampden affair were at their thickest that my earnings invested in railways and mines continued depreciating so constantly as to be a source of great anxiety to me, and every effort to extricate myself by seeking better investments only made matters worse. It was at this time that the endeavour to get the Epping Forest appointment failed, and had it not been for the kindness of a relative, Miss Roberts, of Epsom, a cousin of my mother's, with whose family I had been intimate from my boyhood, I should have been in absolute want. She had intended to leave me £1000 in her will, but instead of doing so transferred it to me at once, and as it was in an excellent security, and brought me in from £50 to £65 a year, it was most welcome. I had sold my house at Grays fairly well, and in 1880 bought a piece of land and built a cottage at Godalming, so that I had a home of my own; but I had now to depend almost entirely on the little my books brought me in, together with a few lectures, reviews, and other articles. I had just finished writing my "Island Life," and had no idea that I should ever write another important book, and I therefore saw no way of increasing my income, which was then barely sufficient to support my family and educate my two children in the most economical way. From this ever-increasing anxiety I was relieved through the grant of a Civil Service Pension of £200, which came upon me as a very joyful surprise. My most intimate and confidential friend at this time was Mrs. Fisher (then Miss Buckley), and to her alone I mentioned my great losses, and my anxiety as to any sure source of income. Shortly afterwards she was visiting Darwin, and mentioned it to him, and he thought that a pension might be granted me in recognition of my scientific work. Huxley most kindly assisted in drawing up the necessary memorial to the Prime Minister, Mr. Gladstone, to whom Darwin wrote personally. He promptly assented, and the next year, 1881, the first payment was made. Other of my scientific friends, I believe, signed the memorial, but it is especially to the three named that I owe this very great relief from anxiety for the remainder of my life.

I have already stated that what at the time appeared to be the great misfortune of the loss of about half of my whole Amazonian collections by the burning of the ship in which I was coming home, was in all probability a blessing in disguise, since it led me to visit the comparatively unknown Malay Archipelago, and, perhaps, also supplied the conditions which led me to think out independently the theory of natural selection. In like manner I am now inclined to see in the almost total loss of the money value of my rich collections, another of those curious indications that our misfortunes are often useful, or even necessary for bringing out our latent powers. I am, and have always been, constitutionally lazy, without any of that fiery energy and intense power of work possessed by such men as Huxley and Charles Kingsley. When I once begin any work in which I am interested, I can go steadily on with it till it is finished, but I need some definite impulse to set me going, and require a good deal of time for reflection while the work is being done. Every important book I have undertaken has been due to an impulse or a suggestion from without. I spent five years in quiet enjoyment of my collections, in attending scientific meetings, and in working out a few problems, before I began to write my "Malay Archipelago," and it was due to the repeated suggestions of my friends that I wrote my "Geographical Distribution of Animals."

But if the entire proceeds of my Malayan collections had been well invested, and I had obtained a secure income of £400 or £500 a year, I think it probable that I should not have written another book, but should have gone to live further in the country, enjoyed my garden and greenhouse (as I always have done), and limited my work to a few lectures and review articles, but to a much less extent than I actually have done. It was the necessity of earning money, owing to my diminishing income, that caused me to accept invitations to lecture, which I always disliked; and the same reason caused me to seek out subjects for scientific or social articles which, without that necessity, would never have been written. Under such conditions as here supposed, my dislike to lecturing would probably have increased, and I should never have ventured on my lecturing tour in America, in which case I should not have written "Darwinism," and, I firmly believe, should not have enjoyed such good health as I am now doing. Then, too, I should probably not have accepted Dr. Lunn's invitation to lecture at Davos, and my two later books would never have come into existence. If, therefore, my books and essays have been of any use to the world—and though I cannot quite understand it, scores of people have written to me telling me so—then the losses and the struggles I have had to go through have been a necessary discipline calculated to bring into action whatever faculties

I possess. I may be allowed here to give an extract from one of these letters on my literary work, nearly the last I received from my lamented friend F. W. H. Myers. He writes (April 12, 1898):—

"I am glad to take this opportunity of telling you something about my relation to one of your books. I write now from bed, having had severe influenzic pneumonia, now going off. For some days my temperature was 105°, and I was very restless at night— anxious to read, but in too sensitive and fastidious a state to tolerate almost any book. I found that almost the only book which I could read was your 'Malay Archipelago.' Of course I had read it before. In spite of my complete ignorance of natural history there was a certain uniqueness of charm about the book, both moral and literary, which made it deeply congenial in those trying hours. You have had few less instructed readers; but very few can have dwelt on that simple, manly record with a more profound sympathy."

Other people, quite strangers, have also told me that they have read it over and over again, and always take it with them on a journey. This is the kind of thing I cannot understand. It is true, if I open it myself I can read a chapter with pleasure; but, then, to me it recalls incidents and feelings almost forgotten, and renews the delights of my wanderings in the wilderness and of my intense interest in the wonderful and beautiful forms of plant, bird, and insect life I was continually meeting with. Others have written in almost equally laudatory terms of my books on "Land Nationalization" and on "Spiritualism," which have introduced them to new spheres of thought; while others, again, have been equally pleased with parts of my "Wonderful Century "and "Man's Place in the Universe." I am thus forced to the conclusion that my books have served to instruct and to give pleasure to a good many readers, and that it is therefore just possible that my life may have been prolonged, and its conditions modified so as to afford the required impulse and the amount of time for me to write them.

CHAPTER XXIII

MY CHARACTER—NEW IDEAS

I HAVE already given an estimate of my character when I came of age. I will now make a few further remarks upon it as modified by my changed views of life, owing to my becoming convinced of the reality of a spirit world and a future state of existence.

Up to middle age, and especially during the first decade after my return from the East, I was so much disinclined to the society of uncongenial and commonplace people that my natural reserve and coldness of manner often amounted, I am afraid, to rudeness. I found it impossible, as I have done all my life, to make conversation with such people, or even to reply politely to their trivial remarks. I therefore often appeared gloomy when I was merely bored. I found it impossible, as some one had said, to tolerate fools gladly; while, owing to my deficient language-faculty, talking without having anything to say, and merely for politeness or to pass the time, was most difficult and disagreeable. Hence I was thought to be proud or conceited. But later on, as I came to see the baneful influence of our wrong system of education and of society, I began to realize that people who could talk of nothing but the trivial amusements of an empty mind were the victims of these social errors, and were often in themselves quite estimable characters.

Later on, when the teachings of spiritualism combined with those of phrenology led me to the conclusion that there were no absolutely bad men or women, that is, none who, by a rational and sympathetic training, and a social system which gave to all absolute equality of opportunity, might not become useful, contented and happy members of society, I became much more tolerant. I learnt also to distrust all first impressions; for I repeatedly came to enjoy the society of people whose appearance or manner had at first repelled me, and even in the most apparently trivial-minded was able to find some common ground of interest or occupation. I feel myself that my character has continuously improved, and that this is owing chiefly to the teaching of spiritualism, that we are in every act and thought of our lives here building up a character which will largely determine our happiness or misery hereafter; and also, that we obtain the greatest happiness ourselves by doing all we can to make those around us happy.

As I have referred in various parts of this volume to ideas, or suggestions, or solutions of biological problems, which I have been the first to put forth, it may be convenient if I here give a brief account of the more important of them, some of which have, I think, been almost entirely overlooked.

1. The first and perhaps the most important of these is my independent discovery of the theory of natural selection in 1858, in my paper on "The Tendency of Varieties to depart indefinitely from the Original Type." This is reprinted in my "Natural Selection and Tropical Nature;" and it has been so fully recognized by Darwin himself and by naturalists generally that I need say no more about it here.

2. In 1864 I published an article on "The Development of Human Races under the Law of Natural Selection," the most original and important part of which was that in which I showed that so soon as man's intellect and physical structure led him to use fire, to make tools, to grow food, to domesticate animals, to use clothing, to build houses, the action of natural selection was diverted from his body to his mind, and thenceforth his physical form remained stable while his mental faculties improved. This paper was greatly admired by Mr. Darwin and several other men of science, who declared it to be entirely new to them; but owing to its having been published in one of my less popular works, "Contributions to the Theory of Natural Selection," it seems to be comparatively little known. Consequently, it still continues to be asserted or suggested that because we have been developed physically from some lower form, so in the future we shall be further developed into a being as different from our present form as we are different from the orang

or the gorilla. My paper shows *why* this will not be; *why* the form and structure of our body is permanent, and that it is really the highest type now possible on the earth. The fact that we have not improved physically over the ancient Greeks, and that most savage races— even some of the lowest in material civilization— possess the human form in its fullest symmetry and perfection, affords evidence that my theory is the true one.

3. In 1867 I gave a provisional solution of the cause of the gay, and even gaudy colours of many caterpillars, which was asked for by Darwin, and which experiment soon proved to be correct. This is fully described in my "Natural Selection and Tropical Nature," pp. 82–86. The principle established in this case has been since found to be widely applicable throughout the animal kingdom.

4. In 1868 I wrote a paper on "A Theory of Birds' Nests," the chief purport of which was to point out and establish a connection between the colours of female birds and the mode of nidification which had not been before noticed. This led to the formulation of the following law, which has been very widely accepted by ornithologists: *When both sexes of birds are conspicuously coloured, the nest conceals the sitting bird; but when the male is conspicuously coloured and the nest is open to view, the female is plainly coloured and inconspicuous.*

No less than fifteen whole families of birds and a number of the genera of other families belong to the first class, of brightly coloured birds with sexes alike, and they all build in holes or make domed nests. Most of these are tropical, but the woodpeckers and kingfishers are European. In the second class, however brilliant the male may be, if the nest is open to view, the female is always plainly coloured, sometimes so much so as to be hardly recognizable as the same species. This is especially the case in such birds as the brilliant South American chatterers and the Eastern pheasants and paradise birds. This law is of especial value, as showing the exceptional need of protection of female birds as well as butterflies, and the remarkable way in which the colours of both classes of animals have become modified in accordance with this necessity. This paper forms chapter vi. of my "Natural Selection and Tropical Nature."

5. In the great subject of the origin, use, and purport of the colours of animals, there are several branches which, I believe, I was the first to call special attention to. The most important of these was the establishment of the class of what I termed "Recognition colours," which are of importance in affording means for the young to find their parents, the sexes each other, and strayed individuals of returning to the group or flock to which they belong. But perhaps even more important is the use of these special markings

or colours during the process of the development of new species adapted to slightly different conditions, by checking intercrossing between them while in process of development. It thus affords an explanation of the almost universal rule, that closely allied species differ in colour or marking even when the external structural differences are exceedingly slight or quite undiscoverable. The same principle also explains the general symmetry in the markings of animals in a state of nature, while under domestication it often disappears: difference of colour or marking on the two sides would render recognition difficult. This principle was first stated in my article on "The Colours of Animals and Sexual Selection" (in "Natural Selection and Tropical Nature," 1878) and more fully developed in "Darwinism." I am now inclined to think that it accounts for more of the variety and beauty in the animal world than any other cause yet discovered.

I may here add that I believe I was first to give adequate reasons for the rejection of Darwin's theory of brilliant male coloration or marking being due to female choice.

6. The general permanence of oceanic and continental areas was first taught by Professor J. D. Dana, the eminent American geologist, and again by Darwin in his "Origin of Species"; but I am, I believe, the only writer who has brought forward a number of other considerations, geographical and physical, which, with those of previous writers, establish

the proposition on almost incontrovertible grounds. My exposition of the subject is given in "Island Life" (chap, vi.), while some additional arguments are given in my "Studies "(vol. i. chap. ii.). The doctrine may be considered as the only solid basis for any general study of the geographical distribution of animals, and it is for this reason that I have made it the subject of my careful consideration.

7. In discussing the causes of glacial epochs I have adopted the general views of Mr. James Croll as to the astronomical causes, but have combined them with geographical changes, and have shown how the latter, even though small in amount, might produce very important results. In particular I have laid stress on the properties of air and water in *equalizing* temperature over the earth, while snow and ice, by their immobility, produce *cumulative* effects; and thus a lowering of temperature of a few degrees may lead to a country being ice-clad which before was ice-free. This is a vital point which is the very essence of the problem of glaciation; yet it has been altogether neglected in the various mathematical or physical theories which have recently been put forward. My own discussion of the problem in chapter viii. Of "Island Life" has never, so far as I know, been controverted, and I still think it constitutes the most complete explanation of the phenomenon yet given.

8. In 1880 I published my "Island Life," and the last chapter but one is "On the Arctic Element in South Temperate Floras," in which I gave a solution of the very remarkable phenomena described by Sir Joseph Hooker in his "Introductory Essay on the Flora of Australia." My explanation is founded on known facts as to the dispersal and distribution of plants, and does not require those enormous changes in the climate of tropical lowlands during the glacial period on which Darwin founded his explanation, and which, I believe, no biologist well acquainted either with the fauna or the flora of the equatorial zone has found it possible to accept.

9. In 1881 I put forth the first idea of mouth-gesture as a factor in the origin of language, in a review of E. P. Tylor's "Anthropology," and in 1895 I extended it into an article in the *Fortnightly Review,* and reprinted it with a few further corrections in my "Studies," under the title, "The Expressiveness of Speech or Mouth-Gesture as a Factor in the Origin of Language." In it I have developed a completely new principle in the theory of the origin of language by showing that every motion of the jaws, lips, and tongue, together with inward or outward breathing, and especially the mute or liquid consonants at the end of words serving to indicate abrupt or continuous motion, have corresponding meanings in so many cases as to show a fundamental connection. I

thus enormously extend the principle of onomatopœia, in the origin of vocal language. As I have been unable to find any reference to this important factor in the origin of language, and as no competent writer has pointed out any fallacy in it, I think I am justified in supposing it to be new and important. Mr. Gladstone informed me that there were many thousands of illustrations of my ideas in Homer.

10. In 1890 I published in the *Fortnightly Review* an article on "Human Selection," and in 1892 (in the Boston *Arena*) one on "Human Progress, Past and Future." These deal with different aspects of the same great problem—the gradual improvement of the race by natural process; and they were also written partly for the purpose of opposing the various artificial processes of selection advocated by several English and American writers. I showed that the only method of advance for us, as for the lower animals, is in some form of natural selection, and that the only mode of natural selection that can act alike on physical, mental, and moral qualities will come into play under a social system which gives equal opportunities of culture, training, leisure, and happiness to every individual. This extension of the principle of natural selection as it acts in the animal world generally is, I believe, quite new, and is by far the most important of the new ideas I have given to the world.

11. In an article on "The Glacial Erosion of Lake Basins" (in the *Fortnightly Review,* December, 1893), I brought together the whole of the evidence bearing upon the question, and adduced a completely new argument for this mode of origin of the valley lakes of glaciated countries. This is founded on their surface and bottom contours, both of which are shown to be such as would necessarily arise from ice-action, while they would not arise from the other alleged mode of origin—unequal elevation or subsidence.

12. In a new edition of "Stanford's Compendium, Australasia," vol. i., when describing the physical and mental characteristics of the Australian aborigines, I stated my belief that they were really a low and perhaps primitive type of the Caucasian race. I further developed the subject in my "Studies," and illustrated it by photographs of Australians and Ainos, of the Veddahs of Ceylon, and of the Khmers of Cambodia—all outlying members of the same great human race. This, I think, is an important simplification in the classification of the races of man.

Bees' cells.—But besides these more important scientific principles or ideas, there are a few minor ones which are of sufficient interest to be briefly mentioned. In the article on the "Bees' Cell" in the "Annals and Magazine of Natural History," I called attention to a circumstance that had been, I think, unnoticed by all previous writers. An immense deal of ingenuity and of mathematical skill had been expended in showing that the two layers of hexagonal cells, with basal dividing-plates inclined at a particular angle, gave the greatest economy of space and of material possible; and the instinct of the bees in building such a comb to contain their store of honey was held to show that it was a divinely bestowed special faculty. But all these writers omitted to take into account one fact, which shows their whole argument to be fallacious. This is, that the combs are suspended *vertically,* and that when full of honey the upper rows of cells have to support at least ten times as much weight as the lowest rows. But there is no corresponding difference in the thickness of the walls of the cells; so that, as the upper rows are strong enough, the lower must be quite unnecessarily strong, and there is thus a great waste of wax. The whole conception of a supernatural faculty for the purpose of economizing wax is thus shown to be fallacious. Darwin's explanation entirely obviates this difficulty, since it depends on the bees possessing intelligence enough to reduce all the cellwalls to a nearly uniform thickness, being that which is sufficient under all circumstances to support the weight of the whole mass of comb and honey.

In an article on "The Problem of Instinct" in my "Studies" (vol. i. chap, xxii.), I have supplemented the usual theory as to *why* birds migrate, by another as to *how* they migrate, and trace it wholly to experience, the young birds following the old ones; but an enormous proportion of the young fail to make the outward or the homeward journey safely.

In 1894 I wrote an article for the *Nineteenth Century* on the question of the proper observance of Sunday, which I have reprinted in my "Studies" under the title, "A Counsel of Perfection for Sabbatarians." In this short article I define clearly, I think for the first time, what the "work" so strictly and impressively forbidden really is, and then show how utterly inconsistent are the great majority of Sabbatarians, who themselves break the commandment both in letter and spirit, while they loudly condemn others for acts which are not forbidden by it. I also show how the commandment can be and should be strictly kept by all who believe it to be a Divine command, and point out the good results which would follow such a mode of obeying it. That the idea was new and its reasoning unanswerable may be perhaps inferred from the fact that no reply, so far as I know, was made to it; while a well-known writer was so impressed by it that he made his own bed the following Sunday in accordance with its suggestions.

CHAPTER XXIV

PREDICTIONS FULFILLED: LATEST HONOURS.

HAVING devoted some space to an account of my various experiences in connection with modern spiritualism, which have, however, been far less extraordinary than those of many of my friends, I may not improperly conclude this record of my life and experience with a statement of a few of the predictions which I have received at different times, and which have been to some extent fulfilled.

In 1870 and the following years several communications in automatic writing were received through a member of my family purporting to be from my brother William, with whom I had lived so many years. In some of these he referred to my disappointments in obtaining employment and to my money losses, always urging me not to trouble myself about my affairs, which would certainly improve; but I was not to be in a hurry. These messages never contained any proofs of identity, and I did not therefore feel much interest in them, and their ultimate fulfilment, though in quite unexpected ways, cannot be considered to be of any great importance.

Some years later, when we were living at Dorking, my

little boy, then five years old, became very delicate, and seemed pining away without any perceptible ailment. At that time I was being treated myself for a chronic complaint by an American medium, in whom I had much confidence; and one day, when in his usual trance, he told me, without any inquiry on my part, that the boy was in danger, and that if we wished to save him we must leave Dorking, go to a more bracing place, and let him be out-of-doors as much as possible and "have the smell of the earth." I then noticed that we were all rather languid without knowing why, and therefore removed in the spring to Croydon, where we all felt stronger, and the boy at once began to get better, and has had fair health ever since.

Some time afterwards I accompanied a lady friend of mine to have a *séance* with the same medium, she being quite unknown to him. Among many other interesting things, he told us that something would happen before very long which would cause us to see less of each other, but would not affect our friendship. We neither of us could guess what that could be, but a year or two later the lady married a very old friend, a widower, whose wife at the time of the prediction was, I think, alive, while he was living in a distant colony without any expressed intention of coming home. After the marriage they went to live in Devonshire, and for some years we only met at very long intervals. These two cases seem to me to be genuine clairvoyance or prediction.

But much more important than the preceding are certain predictions which were made to me in April, 1896, and which have been fulfilled during the succeeding eight years. At that time I was living at Parkstone in rather poor health and subject to chronic asthma, with palpitations and frequent bronchitis, from which I never expected to recover. I had given up lecturing, and had no expectation of ever writing another book, neither had I the least idea of leaving the house I was living in, which I had purchased and enlarged a few years before. It was under these circumstances that a medium I had visited once in London, Madame G——, was staying with friends at Wimborne, and came to see me, and offered to give me a *séance.*. One of her controls, an old Scotch physician, advised me about my health, told me to eat fish, and assured me that I was not coming to their side for some years yet, as I had a good deal of work to do here. The other control, named "Sunshine," an Indian girl, who seemed to be able to get information from many sources, was very positive in her statements. She said, "You won't live here always. You will come out of this hole. You will come more into the world, and do something public for spiritualism." I replied, "You are quite wrong. I shall never leave this house now, and I shall not appear in public again." But she insisted that she was right, and said, "You will see; and when it comes to pass, remember what I told you." She then said, "Fanny [my sister] sends her love. She loved you more than any one in the world." This I knew to be true, though during her life I did not so fully realize it. Then Sunshine gave me her parting words, speaking slowly and distinctly: "The third chapter of your life, and your book, is to come. It can be expressed as Satisfaction, Retrospection, and Work." These three words were spoken very impressively, and I wrote them at once in a small note-book with capital letters, though I had no notion whatever of what they could refer to, and no belief that they would be in any way fulfilled.

Yet two months later the first step in the fulfilment was taken though Dr. Lunn's invitation to give a lecture at Davos, and my acceptance of it, due mainly to the temptation of a week in Switzerland free of cost and with a pleasant party. As already described, this lecture was the starting-point of all my subsequent work. The very next year brought me renewed health and strength to do the work, as already mentioned. Another year passed, and I received a pressing invitation to take the chair and give a short address at the International Congress of Spiritualists in 1898, which I felt myself unable to refuse, and thus, as I had been told I should, I "did something public for spiritualism." Yet another year, and a great desire for life more in the country than at Parkstone (where we were being surrounded by new building operations) led me to join some friends in trying to find a locality for a kind of home-colony of congenial persons; and though the plan was never carried out, it led ultimately to my finding the site on which to build my present house, and thus "get out of that hole," as I had been told by Sunshine that I should do. And now, looking back upon the eight years of renewed health I have enjoyed, and with constant interesting work, how can this be better described than as "the third chapter of my life;" while "Man's Place in the Universe"—a totally new subject for me—may well be termed the "third chapter of my book," that is, of my literary work. Again, this wholesome activity of body and mind, the obtaining a beautiful site where I am surrounded by grass and woodland, and have a splendid view over moor and water to distant hills and the open sea, with abundance of pure air and sunshine, the building of a comfortable house in one of the choicest spots in the whole district—surely all this was well foretold in the one word "Satisfaction." What has chiefly occupied me in this house—an Autobiography extending over three-quarters of a century—is admirably described by the word "Retrospection." And the whole of this process has involved, or been the result of, continuous and pleasurable "Work."

I will only add here that during the whole of this "third chapter of my life" I had entirely forgotten the particular words of the prediction which I had noted down at the time, and was greatly surprised, on referring to them again for the purpose of this chapter, to find how curiously they fitted the subsequent events. Of course it may be said that every one who reaches my age enjoys "retrospection," but that kind of general looking back to the past is very different from the detailed Retrospection I have had to make in searching out the many long-forgotten incidents and details of my varied life, as here recorded; and the Work this has involved, and the Satisfaction I have had in writing it, seem fully to justify the solemn emphasis with which the prediction was made.

Thus far I had written in "My Life," the proofs of which passed out of my hands in September, 1905, and I never contemplated that anything more would happen to me which would add force and completeness to the fulfilment above recorded. But two things of that nature have happened, both totally unexpected, and in themselves very unlikely, and in neither case was my action in any degree influenced by the thought that they did serve to fulfil more exactly what had been predicted. So much was this the case that this point of view did not occur to me till I came to read again the last pages of "My Life" for the purposes of this edition.

I will first mention, however, that the whole period between the publication of the first edition and the completion of the second, has to me been again a long retrospection of what may be termed the best and most active part of my life, in the lengthy preparation of Spruce's "Notes of a Botanist." This period extends from 1849, when I first met him at Para, to the date of his death in 1893. While studying his Journals and letters I recalled, and to some extent visualized, places and persons in South America long forgotten, as well as the pleasant talks and communications I had with him after his return to England. Four years instead of one have thus been occupied in a long course of agreeable Retrospection.

Far more important than this, however, is a circumstance that happened only last year. Early in the spring of last year I was asked by Archdeacon Colley to help him by appearing as a witness in a civil action between himself and Maskelyne, the celebrated conjuror, respecting phenomena occurring with a medium, Dr. Monck. The Archdeacon had challenged Maskelyne to reproduce the phenomena of materialization, *as described by him* in a lecture delivered at Weymouth during the Church Congress, in October, 1905, and agreed to forfeit £1000 if he did so. It happened that I was the only person in England who had witnessed similar phenomena with Monck, as described in chapter XXI., and it was thought that if I saw Maskelyne's imitation I could say how far it was a reproduction of what I had seen with Monck, and what Mr. Colley described. At first I positively refused to go up to London and expose myself to cross-examination in Court, in order to save the Archdeacon from the consequences of his own too impulsive challenge. But repeated and most earnest letters from him, and the statement that his solicitor and counsel both thought the case might go against him without the evidence which I alone could give; and further, that in that case the cause of spiritualism would be greatly damaged, at length prevailed, and I consented to go. I found Maskelyne's performance to be a ludicrous parody of the actual materialization as both I and Archdeacon Colley described it, and explained in Court the exact difference between them, pointing out that all the essential features of the one were omitted in the other, and that, while the actual phenomena were totally inexplicable by normal causes, Maskelyne's imitation could deceive no one at all familiar with such performances. The result of the trial was that the Archdeacon won his cause, and the contempt and ridicule thrown by the opposing counsel on the alleged phenomena were to some extent neutralized by my positive evidence.

During the whole of this event, and the succeeding discussion in the Press, or in my correspondence, I never once thought of any connection with the prediction. But on reading it again now, I see that it fulfils the conditions far more exactly than did my attendance at the International Congress of Spiritualists. For the prediction was—" You will come out of this hole. You will come more into the world, and do something public for spiritualism." In 1898 I had not "come out of the hole," and attending a congress of spiritualists was hardly "doing something public." But giving evidence in a Court of Justice, which was more or less reported all over the kingdom and the world, and which was believed to have saved the reputation of spiritualism on that occasion, was certainly "doing something public" for the cause, and something quite unanticipated by myself or any one for me. But in this very year another event has occurred even less anticipated by myself, and which again accords with the prediction.

In February, 1908, the Linnean Society decided to celebrate the fiftieth anniversary of the reading of the joint Essays of Darwin and myself on July 1, 1858, by a Jubilee Celebration and the presentation of a medal with our two portraits to persons connected with the event; and the Council invited me to attend, to give a short address, and to receive one of the medals.

This wholly unexpected and very unusual honour was in due course conferred upon me in the presence of many of the most eminent naturalists, British and Foreign. Here, then, was yet another "coming more into the world," which I thought I had taken my leave of—the world of Science; and it was an occasion of publicity that could hardly be exceeded, since the record of it would be carried by the Press throughout the whole extent of the civilized world. Here was surely "Satisfaction" for any man, however greedy of praise he might be. For myself, I can only say that I would have been fully content with a lower place than that accorded me, and feel that I attained to the honour more from the accident of my having lived to see the Celebration, than from any idea that I could have the slightest claim to be placed on anything approaching a level with Darwin.

With this celebration, and the publication of this volume, I close my Memoirs. I now wish my readers, who have so far followed the record of my life-history, a hearty Farewell.